General Video Game
Artificial Intelligence

Synthesis Lectures on Games and Computational Intelligence

Editor
Daniel Ashlock, *University of Guelph*

Synthesis Lectures on Games & Computational Intelligence is an innovative resource consisting of 75-150 page books on topics pertaining to digital games, including game playing and game solving algorithms; game design techniques; artificial and computational intelligence techniques for game design, play, and analysis; classical game theory in a digital environment, and automatic content generation for games. The scope includes the topics relevant to conferences like IEEE-CIG, AAAI-AIIDE, DIGRA, and FDG conferences as well as the games special sessions of the WCCI and GECCO conferences.

General Video Game Artificial Intelligence

Diego Pérez Liébana, Simon M. Lucas, Raluca D. Gaina, Julian Togelius, Ahmed Khalifa, and Jialin Liu

ISBN: 978-3-031-00994-5   paperback
ISBN: 978-3-031-02122-0   ebook
ISBN: 978-3-031-00170-3   hardcover

DOI 10.1007/978-3-031-00171-0

A Publication in the Springer series
SYNTHESIS LECTURES ON GAMES AND COMPUTATIONAL INTELLIGENCE

Lecture #5
Series Editor: Daniel Ashlock, *University of Guelph*
Series ISSN
Print 2573-6485   Electronic 2573-6493

General Video Game Artificial Intelligence

Diego Pérez Liébana, Simon M. Lucas, and Raluca D. Gaina
Queen Mary University of London

Julian Togelius and Ahmed Khalifa
New York University

Jialin Liu
Southern University of Science and Technology

SYNTHESIS LECTURES ON GAMES AND COMPUTATIONAL INTELLIGENCE #5

ABSTRACT

Research on general video game playing aims at designing agents or content generators that can perform well in multiple video games, possibly without knowing the game in advance and with little to no specific domain knowledge. The general video game AI framework and competition propose a challenge in which researchers can test their favorite AI methods with a potentially infinite number of games created using the Video Game Description Language. The open-source framework has been used since 2014 for running a challenge. Competitors around the globe submit their best approaches that aim to generalize well across games. Additionally, the framework has been used in AI modules by many higher-education institutions as assignments, or as proposed projects for final year (undergraduate and Master's) students and Ph.D. candidates.

The present book, written by the developers and organizers of the framework, presents the most interesting highlights of the research performed by the authors during these years in this domain. It showcases work on methods to play the games, generators of content, and video game optimization. It also outlines potential further work in an area that offers multiple research directions for the future.

KEYWORDS

computational intelligence, artificial intelligence, video games, general video game playing, GVGAI, video game description language, reinforcement learning, Monte Carlo tree search, rolling horizon evolutionary algorithms, procedural content generation

Contents

6 Procedural Content Generation in GVGAI 97

Ahmed Khalifa and Julian Togelius

7 Automatic General Game Tuning 117

Diego Pérez Liébana

Preface

General video game AI (GVGAI) was originally inspired by one of the previous works carried out at a Dagstuhl seminar in 2013, in which a group of researchers played with the idea of defining a language for video games and building a framework (and competition) around it for agents to play these games. The original ideas outlined in this work (general video game playing, game classification, game tuning and game generation) have materialized over the years in concrete software, general approaches, research projects and an international competition.

The GVGAI framework and competition were born at the University of Essex (UK) in 2014. Previous experience of organizing competitions at this University made the Game Intelligence Group a perfect environment to start. Spyridon Samothrakis, Simon M. Lucas, and Diego Pérez Liébana kick-started then this ambitious project, in collaboration with Julian Togelius (then at ITU Copenhagen, Denmark) and Tom Schaul (then at New York University, U.S.), who developed the first engine for the video game description language (VGDL) in Python. The GVGAI competition was held for the first time in 2014 at the IEEE Conference on Computational Intelligence and Games. Since that first edition, focused on planning agents playing a total of 30 different games, the *GVGAI world* has kept growing. On July 2019, the framework counts on 200 games and 5 different competition tracks. Raluca D. Gaina joined the team to bring the 2-Player track, Ahmed Khalifa brought the Procedural Content Generation challenges and Jialin Liu led the (repeatedly requested) learning track. Since 2014, more than 200 students and researchers around the globe have submitted agents and generators to the challenge, and several high-level education institutions have used GVGAI to propose assignments and student projects at different levels (undergraduate, master and Ph.D.). A recent survey on GVGAI cites around 100 papers that use GVGAI for research.

This book is a summary of our work on GVGAI for the last five years. This book presents the main components of GVGAI, from the video game description language (VGDL) used to generate games to the framework itself and the multiple opportunities for research it presents. The book is also a collection of what is, in our opinion, our most relevant work on this domain and dives deep into optimization and control algorithms. Finally, multiple exercises and project ideas are proposed at the end of each chapter as suggestions to take this research further.

Diego Pérez Liébana, Simon M. Lucas, Raluca D. Gaina, Julian Togelius, Ahmed Khalifa, and Jialin Liu
August 2019

Acknowledgments

During these five years of GVGAI work, we have had the pleasure to collaborate with many people on this challenge. A special mention needs to be made to our multiple co-authors during these years. Spyridon Samothrakis and Tom Schaul, members of the GVGAI steering committee, played a crucial role defining the shape of the challenge and competition at the start. Philip Bontrager and Ruben Torrado implemented the GVGAI-Gym version of the framework for the learning track of the competition. Sanaz Mostaghim's expertise on multi-objective Optimization and Kamolwan Kunanusont's work on game optimisation also contributed to the research presented in this book.

We would also like to thank Adrien Couëtoux, Dennis Soemers, Mark Winands, Tom Vodopivec, Florian Kirchgessner, Jerry Lee, Chong-U Lim and Tommy Thompson. They built the first set of GVGAI agents that were used for testing the framework, both the single and the two-player planning tracks of the competition. Thanks also go to Cameron Browne and Clare Bates Congdon for their read, review and useful comments made to improve this book.

Finally, we would also like to thank the Game AI community that has shown interest in this challenge. Thanks to the members of the group at the 2013 Dagstuhl seminar on "Artificial and Computational Intelligence in Games," the instructors that have used GVGAI for setting assignments for their modules and the participants that have submitted a controller or a generator to the competition.

Diego Pérez Liébana, Simon M. Lucas, Raluca D. Gaina, Julian Togelius, Ahmed Khalifa, and Jialin Liu
August 2019

CHAPTER 1

Introduction

Diego Pérez Liébana

The general video game AI (GVGAI) framework and its associated competition have provided to the AI research community with a tool to investigate General AI in the domain of games. For decades, games have been used as benchmarks to perform AI research: they are fast and cheap simulations of the real world and (this can't be overlooked) they are fun. Soon thereafter, work in games led to establishing comparison of AI performance among researchers, which subsequently took investigators and practitioners to establish competitions around them.

Game-based AI competitions are a very popular and fun way of benchmarking methods, becoming one of the most central components of the field. Research around competitions provides a common benchmark in which researchers can try their methods in a controlled, fair and reliable manner, opening the possibility of comparing multiple approaches within exactly the same environment. Using games for competitions allows researchers to show the results in a way most people can understand and interpret, in a domain that is typically challenging for humans as well, helping raise awareness of AI research among different groups and communities. Either as vehicles for creating game playing agents or generation of content, participants submit controllers or generators which are evaluated in the same conditions to proclaim a winner.

Without any doubt, advances in research have been achieved when working in one game only (such as Transposition Tables and Iterative Deepening from Chess, or Monte Carlo Tree Search from Go), but it is also true that an agent that performs well at one of these contests would likely not perform above chance level in any of the others (even if the same interface was provided). The initial thought that an AI that beats humans at Chess would achieve general intelligence needed to be evaluated again: single-game competitions have the drawback of overfitting a solution to the game used. Then, the value of the competition as a benchmark for general AI methods is limited: participants will tend to tailor their approaches to the game to improve their results. The results of the competition can therefore show the ability of the contestants to engineer a domain specific solution rather than the quality of the AI algorithm that drives the competition entry. For example, participants of the Simulated Car Racing Competition have gotten better at lap times during the more than five years this contest took place, but the presence of game-specific engineering arguably made the approaches less general. Transferring the algorithms and learning outcomes from this competition to other domains has become more and more difficult. It is sensible to think that a competition designed with this in mind would allow us to learn more about artificial *general* intelligence.

The present book details the efforts of GVGAI as a framework (and competition) designed with the intention of promoting research focused on not one but many games. This is done in a series of different chapters that cover the distinct angles of GVGAI.

- Chapter 1 (the one you are reading) introduces GVGAI, from its predecessors until its origins, and provides a vision of the uses of this framework and competition in education, as well as the potential impact it may have in the games industry.

- Chapter 2 describes the Video Game Description Language (VGDL) used to implement games and levels, as well as how the GVGAI framework works and its associated competitions are organized.

- The first one of the more technical chapters is number 3, which focuses on the planning problem tackled in GVGAI. In this case, agents are asked to play any game is given with the aid of a forward model (FM), a simulator that allows agents to play-test moves during their thinking time. The main two methods that this chapter presents are Monte Carlo tree search (MCTS) and rolling horizon evolutionary algorithms (RHEA), as well as several enhancements of these.

- The planning problem proposed for GVGAI is the one that has received more attention. Due to this, Chapter 4 introduces the state-of-the-art controllers that have achieved better results in the competition during the years. It also discusses what are the main hazards and open problems of this challenge, to then propose avenues of future work in this area.

- Chapter 5 discusses the use of GVGAI in a learning setting. In this case, the game playing agents must learn how to play a given game by repeatedly playing the same game during a training phase, this time without the aid of an FM. This chapter describes the particularities of the framework for this end, as well as its integration with Open AI Gym, facilitating the use of reinforcement learning (RL) algorithms for this challenge.

- GVGAI has also applications in procedural content generation (PCG). The framework and competition allow the implementation of rule and level generators. Chapter 6 describes these two aspects of the framework, explaining how can these generators be created and what are the current approaches in this area.

- Chapter 7 focuses on Automatic Game Tuning within GVGAI. This chapter explores how VGDL games can be parameterized to form a *game space*: a collection of possible variants of the same game that could include hidden gems or a perfect gameplay balance from a design point of view. An evolutionary algorithm (the n-tuple bandit evolutionary algorithm; NTBEA) is presented and used to tweak variables of several games in order to obtain variants that serve a pre-determined purpose.

- Chapter 8 introduces a new concept to GVGAI, which is the usage of games and AI agents described in a high-level language that are compatible with games and agents from GVGAI. This chapter describes what is needed to have games that support an FM in an efficient way, as well as presenting a common interface to facilitate the compatibility between systems, illustrated with concrete examples.

- Finally, Chapter 9 concludes this book by asking "What's next?," outlining outstanding research with the current challenge, but also the possibilities of the framework and competition to keep proposing new and interesting problems.

This book brings an overall perspective on the motivation, framework, competition and related research work that forms the GVGAI world. It also provides insights about how the framework has been used in Education during these years and how it can be useful for the Games Industry. Last but not least, each technical chapter proposes, at the end, a series of exercises for those who want to dive deeper in this framework and the challenges it proposes. All these exercises can also be found at the book's website: `https://gaigresearch.github.io/gvgaibook/`.

1.1   A HISTORICAL VIEW: FROM CHESS TO GVGAI

The utilization of games as benchmarks for AI research is as old as the field itself. Alan Turing proposed Chess as an AI benchmark and studied how to play this game with a manual execution of Minimax [Turing, 1953]. Inaugurated in the 1970s, the World Computer Chess Championship has systematically compared AI algorithms in this game until this day [Newborn, 2003]. Since IBM's *Deep Blue* defeated the best human player at its time, Garry Kasparov, the championship has continued pitting computer against computer in a chess AI arms race. Nowadays, we are living through the same history again with the game Go, after Deepmind's *AlphaGo* [Silver et al., 2016] beat the World Go Champion Lee Sedol in 2017, or the first Starcraft II professionals in 2019 [Deepmind, 2019].

Since the start of the 21st century, many competitions and benchmarks for video games have come to existence. The nature of the games that form the basis of these competitions is incredibly varied. Some competitions focus on first-person shooter games, such as Unreal Tournament 2004 [Hingston, 2010] and VizDoom [Kempka et al., 2016]; platform games such as Super Mario Bros [Togelius et al., 2013] and Geometry Friends [Prada, Melo, and Quiterio, 2014]; car racing, such as TORCS [Loiacono et al., 2010]; classic arcade games, such as Ms. Pac-Man [Rohlfshagen and Lucas, 2011] real-time strategy games, such as StarCraft [Ontanon et al., 2013] and Starcraft II [Vinyals et al., 2017]; and navigation problems such as the Physical TSP game and competition [Pérez Liébana, Rohlfshagen, and Lucas, 2012]. The large majority of the existent competitions, however, focus on a single game.

In the study of artificial general intelligence (AGI), a predominant idea is that the intelligence of an agent can be measured by its performance across a large number of environments.

Schaul, Togelius, and Schmidhuber [2011] formalized this idea by arguing that these environments could be games sampled from a game space. Game spaces can be defined via a game description language (GDL) and an engine that parses and executes it in a playable way.

Many authors have developed GDLs to define at different levels of abstraction. Examples are the Ludi language for combinatorial board games [Browne and Maire, 2010] and the Card Game Description Language [Font et al., 2013]. An overview of several GDLs and representations for game rules and mechanics can be found in Nelson et al. [2014].

However, the first attempt at finding a remedy to this problem via a GDL was done by the Stanford University Logic Group in 2005 [Genesereth, Love, and Pell, 2005], when they ran the first *General Game Playing Competition* (GGP). In this contest, competitors submit agents that can play a number of previously unseen games, usually turn-based discrete games or variants of existing board games. Games were described in a GDL [Love et al., 2008] and could be single or two-player deterministic games (Figure 1.1 shows a few examples of existing games). GDL is based on first-order logic (syntactically, it's a version of Prolog), rather low-level. For instance, the description of a game like Tic-Tac-Toe is several pages long. GGP agents have approximately 1 sec of thinking time to make a move and the description of the game is given to the players so it can be analyzed.

Figure 1.1: Example games from the General Game Playing Competition.

Another popular environment is the arcade learning environment (ALE), which is based on an emulator of the Atari 2600 game console [Bellemare et al., 2013]. This collection of video games include simple-looking (but some of very high quality, such as Pac-Man or Space Invaders) games. Two of these games can be seen in Figure 1.2.

Agents developed for playing ALE games are meant to be able to play any game within this framework in real-time, providing actions every few milliseconds. In recent years, research with this framework has become increasingly popular, especially using deep learning to train agents that receive the raw screen capture as input, plus a score counter [Mnih et al., 2015].

In most cases, the algorithms learn to play one ALE game at a time, having to train again when the environment is different and forgetting what was learned for the first game. Multi-

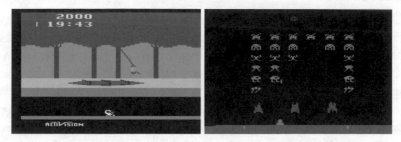

Figure 1.2: Example games from the Arcade Learning Environment: *Pitfall* (left) and *Space Invaders* (right).

task and transfer learning are one of the main focuses of research in this type of problems at the moment [Weiss, Khoshgoftaar, and Wang, 2016]. Due to the nature of the Atari emulator, all games in ALE are deterministic. This limits the adaptability of the methods trained to play these games and makes them too brittle when facing changes in the environment. Additionally, games are loaded via ROMs instead of being defined in a GDL, so they can't be modified easily.

Another fruitful area of research for multi-task learning is that of multi-agent. This domain does not only have the challenge of learning in many games at the same time, but also dealing with opponent modeling. The Multi-Agent Reinforcement Learning in MalmÖ (MARLÖ) framework and competition is a new challenge that proposes this type of research in the game Minecraft [Pérez Liébana et al., 2018a]. The MARLÖ competition ran for the first time in November/December 2018 and it proposes multi-task, multi-agent learning via screen capture through the OpenAI Gym interface. Figure 1.3 shows the three games that were used in this challenge.

Figure 1.3: MARLÖ 2018 Competition games. From left to right, *Mob Chase* (two players collaborate to catch a mob in a pin), *Build Battle* (two players compete to build a structure that matches a model) and *Treasure Hunt* (where two players collaborate in collecting treasures in a maze while avoiding enemy zombies).

MARLÖ proposes a similar task to the one tackled in ALE, but stepping up the game by using 3D environments and a more general approach. Each one of these games can be parameterized, so blocks, objects and character appearance can vary from instance to instance (also known as *tasks*). Game settings and tasks are described in XML files and the final rankings of this challenge are computed in task *a priori* unknown by the participants.

Ebner et al. [2013] and Levine et al. [2013] described the need and interest for a framework that would accommodate a competition for researchers to tackle the challenge of general video game playing (GVGP). The authors proposed the design of the video game description language (VGDL), which was later developed by Schaul [2013a, 2014] in the Python framework py-vgdl. VGDL was designed so that it would be easy to create games both for humans and algorithms, eventually allowing for automated generation of test-bed games. VGDL describes real-time arcade games with stochastic effects, hidden information and played by an avatar. VGDL is a language that allows the creation of a potentially infinite number of games in a very compact manner, describing them in fewer lines than GGP's GDL and MARLÖ's XML format.

VGDL is the base of the GVGAI Framework and Competition, subject of this book. GVGAI was developed as a new iteration of py-vgdl in Java, exposing an API for agents to play games defined in VGDL. Researchers should develop their agents without knowing which games would they be playing. A competition server was made be available so participants could submit their bots and be evaluated in an unseen set of games. In order to make games more appealing to human players, a special care was put into providing a nice set of graphics and sprites within the framework. One of the still standing goals of GVGAI is to be able to compare bot and human performance, and even creating environments in which human and AIs can collaborate and compete in GVGP. Figure 1.4 shows the evolution of one of the VGDL/GVGAI games, Pac-Man, through the different versions of the framework.

Initially, GVGAI was mainly focused on proposing a competition for single-player game agents, in which controllers have access to the model of the game (albeit not the VGDL description) promoting research in model-based RL algorithms. Shortly after, VGDL was expanded to account for 2-player games, adding the challenge of opponent player modeling in GVGP planning. The learning track of the GVGAI competition completes the available settings for general game playing agents. In this last scenario, no FM is provided, challenging model-free RL methods that would learn from static game observations, screen capture, or both.

The framework was also expanded to exploit the versatility of VGDL by proposing challenges on procedural content generation (PCG). Two extra tracks dedicated to the generation of game rules and levels for any game within the framework have been recently created. Another potential use of GVGAI is for game prototyping, and there is a growing body of research using the framework for game design and the implementation of mixed-initiative design tools.

---

[1]Oryx Design Lab, https://www.oryxdesignlab.com/.

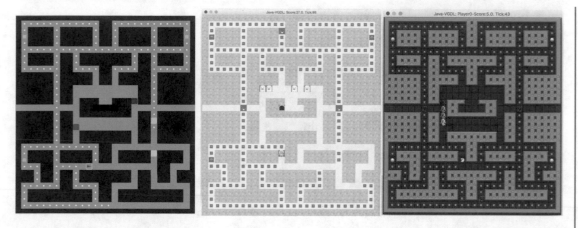

Figure 1.4: Evolution of Pac-Man in py-vgdl and GVGAI frameworks. From left to right, Pac-Man in py-vgdl (with colored squares), initial (with simple sprites) and current GVGAI framework (with sprites made by a graphic designer,[1] with support for transparencies, animations and background auto-tiling).

GVGAI has been used in multiple research domains, and this is the focus of the present book. Nevertheless, GVGAI has had an important impact in Education, being used in undergraduate and master taught modules as assignments, final year undergraduate and master projects and Ph.D. topics. We also strongly believe that GVGAI can also have a great impact in the games industry. We conclude this chapter highlighting these impact cases, before diving into the research usages of the framework and competition.

1.2   GVGAI IN EDUCATION

The GVGAI framework has been employed by instructors from many institutions around the globe to set up engaging assignments in their taught modules. Projects around the GVGAI framework and competition have been proposed for undergraduate and master thesis and Ph.D. topics. This section describes some of these usages, although it does not intend to be an exhaustive list.

The GVGAI framework has been used in at least two different ways in taught modules. The most common usage of the framework is around the game playing challenge for single and 2-player games. Students are first taught of the main GVGAI concepts to then explore how AI algorithms can be implemented in an assignment. These agents, sometimes developed individually and others in groups, participate in either a private league or in the overall competition. The assignment's mark can include how well the entry of each student or group performs in the mentioned league. The following is a not-exclusive list of the institutions that, in the knowledge of the authors, have used the GVGAI framework in their taught modules:

- Otto Von Guericke Universität, Magdeburg, Germany.

- University of Essex, Colchester, United Kingdom.

- University of Muenster, Muenster, Germany.

- Universidad Carlos III de Madrid, Madrid, Spain.

- Universidad de Málaga, Málaga, Spain.

- New York University, New York, United States.

- Southern University of Science and Technology, Shenzhen, China.

- Nanjing University, Nanjing, China.

Some of these institutions have run private leagues (via the GVGAI website and server[2]). In a private league, the module supervisor has full control over how the league is set up, including when and how students can submit entries, how are these evaluated, the set of games that are made available for the competition and, in the 2-player track, the available opponents within the league.

The game design and development capabilities of GVGAI have also been used in taught modules. In this case, VGDL has been explained as a high-level language to create games in. The objective of the assignments is the design and development of interesting games, either manually or guided by AI methods. Examples are creation of new puzzle games or exploring the parameter space of VGDL games (see Chapter 7). Some of the games created in this module have joined the corpus of VGDL games in the GVGAI framework. Examples of higher education institutions that have used the framework in this way are:

- IT University of Copenhagen, Copenhagen, Denmark.

- University of Essex, Colchester, United Kingdom.

- Queen Mary University of London, London, United Kingdom.

Some of the interesting research challenges GVGAI offers have been addressed in the form of master dissertation projects. Most of these have focused on the planning tasks (single and 2-player), which is not surprising given that this is the first track that became available. The GVGAI framework includes sample agents for all tracks, providing an ideal starting point for these projects. These can be used as baselines for comparison and/or as a starting point for algorithmic improvements. Our experience shows that this usage tends to provide an excellent educational experience for the student.

An illustrative example is the work by Maarten de Waard on MCTS with options, which showed how the use of options in MCTS outperformed the vanilla method in most of the

---

[2]www.gvgai.net

games studied. This work started as an master project and was later published as a conference paper [Waard, Roijers, and Bakkes, 2016]. Other uses of GVGAI games in a planning setting for master thesis GVGAI include other enhancements of MCTS [Schuster, 2015], real-time enhancements [Soemers, 2016], MCTS knowledge-based[3] improvements [van Eeden, 2015] and goal-oriented approaches [Ross, 2014].

Episodic learning from screen capture was the focus of Kunanusont's master thesis [Kunanusont, 2016] and the level generation track was the topic of two master projects: Neufeld [2016], who applied Answer Set Programming and was also published in a relevant conference as a paper [Neufeld, Mostaghim, and Pérez Liébana, 2015]; and Nichols [2016], who used genetic algorithms for the first level generation competition. Last but not least, the master project by Gaina (one of the co-authors of this book) [Gaina, 2016] expanded the framework itself to incorporate 2-player games and run the first 2-player GVGAI competition [Gaina et al., 2017a]. This original work was also published as a paper in a conference [Gaina, Pérez Liébana, and Lucas, 2016].

Running the GVGAI competition (and others before) for several years and using the framework in our own taught modules has shown us that accessibility to a framework, documentation and the competitive element of competitions can motivate students to learn and propose new ideas for the problems tackled. The objective of this book is, apart from bringing the latest and most relevant research on GVGAI, to provide a resource for AI module supervisors with examples of possible projects for their students. As such, every technical chapter of this book concludes with a list of proposed exercises for students to attempt. Some of them can be seen as practical tasks, while others can spark new ideas for research projects at different levels.

1.3 GVGAI AND THE GAMES INDUSTRY

The original Dagstuhl report on General Video Game Playing stated that GVGAI games would have their *"unique story and goal which should be achieved by the player. The method of how the player interacts with the game is the same across all these games [...]. This will allow us to gain new insights in working toward the holy grail of artificial intelligence, i.e., development of human-like intelligence"* [Levine et al., 2013]. Although this may sound like a primarily research focused statement, we claim that research on General Video Game Playing can be fruitful for the games industry.

Initially, it is sensible to think that no company would include in their games in development a general agent that can play any game at an average performance level when they can include ad-hoc bots that fit perfectly the environment they are supposed to be immersed in. General agents do not necessarily address certain characteristics that these bots should have.

Namely, AI agents in games need to behave in a `specific` manner: they need to accomplish a certain task and do it at the level it is required to in order to provide a good player experience. They also need to be `predictable`: within reason, agents should behave in a way they

---

[3]See Chapter 3 for the Knowledge-Based MCTS method this work is based on.

are designed to. For instance, games where different types of enemies interact with the player in different ways should be consistent through the lifetime of the game according to certain design specifications. The player would not understand how a particular character that behaves, for example, recklessly during most of the game, exhibits a cautious and planned behavior without any explainable reason for it. Finally, AI agents must perform their duties in an `efficient` manner. For this, they typically use scripted behavior, objectives given by a designer and data structures or object representation built for the game being developed. A general agent that aims not to use these resources may not perform as efficiently as one that does, both in terms of computational cost and actual efficacy of the task at hand.

Before you decide to throw this book to the bin, let us stop here and highlight why we think research on GVGP can be beneficial for the games industry. One of the main goals of GVGP (and, by extension, GVGAI) is to advance the state of the art on general algorithms. For instance, this research aims to improve the capabilities of algorithms that can work in *any* game, such as MCTS and RHEA. The vanilla versions of these methods, as used in GVGAI (see Chapter 3), do not incorporate any domain knowledge of the game. This means the algorithms do not have information about the goal of the game, the best strategy to win, not even details about what the game is about or what do other entities in it do. Therefore, we are aiming at improving the algorithm *itself*, not how it works in one particular game. We are focused on deviating the attention from algorithmic improvements in one game to advances for a broader set of domains. If improving on the reasoning and search capabilities of an algorithm is beneficial to *all* games, then it must be a step forward for *any* game. By doing this, GVGAI aims to provide a better starting point for those programmers interested in taking MCTS and adapt it for their particular game(s).

It is important to highlight that, when we refer to agents that play games, we may not necessarily aim at agents that play those games to win. For instance, agents could be trained to assist the player in any game within the 2-player GVGAI context, in order to identify behaviors for non-player characters (NPC) via opponent modeling. But what if the agents are trained to do other things in any game? For instance, a general agent could be aiming at exploring the environment, or interacting with it, or even finding new and unexpected ways of solving a puzzle or a level. Or maybe you could have a system of agents in which each one of them is able to play with a different role [Guerrero-Romero, Lucas, and Pérez Liébana, 2018]. A system of general agents that can do this could be used to perform automatic play-testing and Q/A for *any* game.

We are convinced that, while reading these lines, you are thinking of a particular game in which this can be applied. It is quite likely that the game you are thinking of is different to the one other readers have in mind. These ideas still apply, and this is the case because the concepts mentioned here are *general*. Could we build such system that works in many games? Could you then adapt it to work in *your* game?

One of the many different configurations that an agent could have is to play as closely as possible as a human would. Creating believable non-player characters is one of the standing challenges in games research and development [Hingston, 2012]. This is a standing problem for not just one, but multiple games. Research can be focused on what are the concepts that make bots believable in general. We believe some of these concepts cut across different games, certainly in the same genre. Researching how an agent can be believable in a group of games will advance the state of the art and provide valuable insights into how to make believable bots for particular games.

As mentioned above, these general agents can be used for automatic play-testing. There is only one step from here to build procedural content generation (PCG) systems. If agents can evaluate games by playing them, they can be incorporated into an automatic system that generates new content. This approach, known as relative algorithm performance profiles (RAPP), is a simulated-based approach very common in the literature [Shaker, Togelius, and Nelson, 2016]. Generality in AI agents allows for generality in content generation. A system that is able to generate content (be this levels, mazes, items, weapon systems, etc.) for any game is given would be rich enough to be adapted to any particular game and provide a good starting point.

GVGAI takes on the ambitious goal of doing research for many games and from many different angles. We hypothesize that, by taking a step away from working on specific games, we are actually improving the state of the art in a way that is more valuable for more researchers and developers.

ACRONYMS

ALE	Arcade Learning Environment
AGI	Artificial General Intelligence
BFS	Best First Search
EOS	End Of Screen
FI2Pop	Feasible Infeasible 2-Population
FM	Forward Model
GDL	Game Description Language
GGP	General Game Playing
GVGAI	General Video Game Artificial Intelligence
GVGP	General Video Game Playing
HOLOP	Hierarchical Open-Loop Optimistic Planning
JSON	JavaScript Object Notation
KB-MCTS	Knowledge-based Fast Evolutionary MCTS
MARLÖ	Multi-Agent Reinforcement Learning in MalmÖ
MCTS	Monte Carlo Tree Search
ML	Machine Learning
NPC	Non-Player Character
NTBEA	N-Tuple Bandit Evolutionary Algorithm
OLETS	Open Loop Expectimax Tree Search
PCG	Procedural Content Generation
RAPP	Algorithm Performance Profiles
RHEA	Rolling Horizon Evolutionary Algorithms
RMHC	Random Mutation Hill Climber
SE	Shannon Entropy
SFP	Statistical Forward Planning
UCB	Upper Confidence Bounds
UCT	Upper Confidence Bounds for Trees
VGDL	Video Game Description Language

CHAPTER 2

VGDL and the GVGAI Framework

Diego Pérez Liébana

2.1 INTRODUCTION

The present chapter describes the core of the framework: the video game description language (VGDL) and the general video game AI (GVGAI) benchmark. All games in GVGAI are expressed via VGDL, a text based description language initially developed by Tom Schaul [Schaul, 2013a]. Having the description of games and levels independent from the engine that runs them brings some advantages. One of them is the fact that it is possible to design a language fit for purpose, which can be concise and expressive enough in which many games can be developed in. The vivid example of this is the 200 games created in this language since its inception. Another interesting advantage of this separation is that it is not necessary to compile the framework again for every new game is created, which helps the sustainability of the software. The engine can be kept in a compiled binary form (or a .jar file) that simply takes games and levels as input data.

But arguably one of the main benefits of having VGDL is the possibility of easily customize games and levels. Its simplicity allows researchers and game designers to create variants of present games within minutes. Put in comparison with other game research frameworks, it is much easier to tweak a VGDL game than modifying an Arcade Learning Environment (ALE) or Unity-ML one—in fact in some cases this may not even be possible. This feature is more than a simple convenience: it opens different research avenues. For instance, it facilitates the creation of a portfolio of games to avoid overfitting in learning agents [Justesen et al., 2018], as well as enabling research on automatic game tuning and balancing (see Chapter 7) and game and level generation (Chapter 6).

Last but not least, the versatility of VGDL eases the organization of a competition that provides new challenges on each one of its editions. Competitions have been part of the game AI community for many years, normally focusing on a single game at a time. As mentioned in the previous chapter, using more games at once allows research to focus more on the generality of the methods and less in the specifics of each game. However, this is not easy to achieve without the capacity of generating games of a medium complexity in a relatively fast and easy way.

Section 2.2 of this chapter describes the VGDL in detail, with special emphasis on its syntax and structure, also providing illustrative examples for single-player and two-player games, using both grid and continuous physics. Section 2.3 describes the framework that interprets VGDL and exposes an API for different types of agents to play the available games. This is followed, in Section 2.4, by a description of the competition and the different editions run over the years.

2.2   THE VIDEO GAME DESCRIPTION LANGUAGE

VGDL is a language built to design 2-dimensional (2D) arcade-type games. The language is designed for the creation of games around objects (*sprites*), one of which represents the player (*avatar*). These objects have certain properties and behaviors and are physically located in a 2D rectangular space. The sprites are able to interact among each other in pairs and the outcomes of these interactions form the dynamics and termination conditions of the game.

VGDL separates the logic into two different components: the game and the level description, both defined in plain text files. All VGDL game description files have four blocks of instructions.

- **Sprite Set:** This set defines the sprites that are available in the game. Each object or sprite is defined with a class and a set of properties. The are organized in a tree, so a child sprite inherits the properties of its ancestors. All types and properties are defined within an ontology, so each sprite corresponds to a class defined in a game engine (in this book, this engine is GVGAI).

- **Interaction Set:** These instructions define the events that happen when two of the defined sprites collide with each other. As in the previous case, the possible effects are defined in an ontology and can take parameters. Interactions are triggered in the order they appear in this definition.

- **Termination Set:** This set defines the conditions for the game to end. These conditions are checked in the order specified in this set.

- **Level Mapping:** This set establishes a relationship between the characters in the level file and the sprites defined in the Sprite Set.

The code in Listing 2.1 shows the complete description of one of the games in the framework: *Aliens*. This game is a version of the traditional arcade game *Space Invaders*, in which the player (situated in the bottom part of the screen) must destroy all aliens that come from above. In the following we analyze the code that implements this game.

In the first line, the keyword `BasicGame` marks the start of the definition of a standard VGDL game (Chapter 7 will study other types of games). In this case, it specifies a parameter, `square_size`, that determines the size in pixels of each cell in the game board. This is purely an aesthetic aspect.

```
 1  BasicGame square_size=32
 2   SpriteSet
 3    background > Immovable img=oryx/space1 hidden=True
 4    base      > Immovable img=oryx/planet
 5    avatar    > FlakAvatar stype=sam img=oryx/spaceship1
 6    missile > Missile
 7      sam   > orientation=UP singleton=True img=oryx/bullet1
 8      bomb  > orientation=DOWN speed=0.5 img=oryx/bullet2
 9    alien   > Bomber stype=bomb prob=0.01 cooldown=3 speed=0.8
10      alienGreen > img=oryx/alien3
11      alienBlue > img=oryx/alien1
12    portal  > invisible=True hidden=True
13      portalSlow > SpawnPoint stype=alienBlue cooldown=16 total=20
14      portalFast > SpawnPoint stype=alienGreen cooldown=12 total=20
15
16   InteractionSet
17    avatar  EOS > stepBack
18    alien   EOS > turnAround
19    missile EOS > killSprite
20
21    base bomb > killBoth
22    base sam > killBoth scoreChange=1
23
24    base    alien > killSprite
25    avatar alien bomb > killSprite scoreChange=-1
26    alien   sam   > killSprite scoreChange=2
27
28   TerminationSet
29    SpriteCounter       stype=avatar                limit=0 win=False
30    MultiSpriteCounter  stype1=portal stype2=alien limit=0 win=True
31
32   LevelMapping
33    . > background
34    0 > background base
35    1 > background portalSlow
36    2 > background portalFast
37    A > background avatar
```

Listing 2.1: VGDL Definition of the game *Aliens*, inspired by the classic game *Space Invaders*.

Lines 2–14 define the **Sprite Set**. There are six different types of sprites defined for *Aliens* (*background*, *base*, *avatar*, *missile*, *alien*, and *portal*), some of them with sub-types. The class that corresponds to each sprite is the keyword that starts with an upper case letter on each type or sub-type. For instance, the *avatar* sprite is of type MovingAvatar while the two types of portals (*portalSlow* and *portalFast*) will be instances of SpawnPoint. As can be seen, the definition of the tpye can be either in the parent of a hierarchy (as in *alien*, line 9) or in the children (as in *portal*, lines 13 and 14).

Some parameters that these sprites receive are common to all sprites. Examples in *Aliens* are img, which describes a file that will graphically represent the sprite in the game, or

Singleton, which will indicate to the engine that only one instance of this sprite can be present in the game at all times. In this game, this sprite is sam (the bullet that the player shoots): in the original game Space Invaders, the player can only shoot one bullet at a time. Another interesting parameter is hidden, which indicates to the engine that the sprite must not be included in the observations the agent receives (see Section 2.3). This is used throughout the games in the GVGAI framework to model partial observability in games. A different parameter, invisible, only hides the sprite in the visual representation of the game, but it is still accessible in the game observations.

Many parameters, however, are dependent on the type of sprite. For instance, the avatar for *Aliens* is set up as a FlakAvatar, which the ontology defines as an avatar type (i.e., controlled by the player) that can move only LEFT and RIGHT and spawn an object, when playing the action USE, of the type defined in the parameter stype. In this case, stype has the sprite *sam* as value, which is defined below in the code.

sam has *missile* as a parent sprite. *missile* is of type Missile, which is a type of sprite that moves in a straight direction ad infinitum. In *Aliens*, there are two types of missiles: *sams* (shot by the player) and *bombs* (dropped by the aliens), which have different parametersations of the type Missile. One of these parameters is orientation, which determines the direction in which the sprite travels. Another one, speed, indicates how many cells per frame the sprite moves. As can be seen in lines 7 and 8, these missiles have different orientations and speeds (if the parameter speed is not defined, it takes the default value of 1.0).

The enemies the player must destroy are described in lines 9–11. *alien* is the parent sprite of a hierarchy that defines them as being of type Bomber. Bombers are sprites that move in a certain direction while spawning other sprites. As in the case of the missiles, their movement is defined by a speed and an orientation (which by default is set to RIGHT). Besides, the parameter cooldown specifies the rate at which the sprites execute an action (in the example, aliens move or shoot once every 3 frames). The parameter stype describes the sprite type that is spawned with a probability (per action) prob as defined in line 9.

Aliens are spawned by portals (lines 13–14), which can be seen as immovable bombers. They can define the maximum number of sprites to spawn, after which the spawner is destroyed, using the parameter total. In this case, probability prob is set by default to 1.0, so the rate at which aliens are spawned is set only by the cooldown parameter.

The Sprite Set is arguably the most complex and richest set in VGDL, but defining complex sprites without the rules that describe the interactions between them does not constitute a game. The **Interaction Set**, lines 16–26, describes this. Each line in the interaction set describes an effect that takes place when two sprites collide with each other. In order to keep the description of the game short, it is possible to group pairs of interactions in instructions with two or more sprites (see line 25). An instruction of the type $A_1 A_2 A_3 \ldots A_N > E$ is equivalent to N effects $A_1 A_2 > E$, $A_1 A_3 > E \ldots A_1 A_N > E$, etc.

The first three interactions (lines 17–19) describe the effects to apply when sprites leave the game screen (keyword EOS—End of Screen). When an avatar reaches the end of the screen, the game will automatically execute the event stepBack, which sets the position of the sprite to its previous location (hence preventing it to leave the screen). When aliens reach the end, however, the event executed is turnAround, which moves the sprite one cell down and flips its orientation. This makes the enemies descend in the level while keeping their lateral movement. Finally, when missiles reach the EOS, they are destroyed. Note this interaction affects both subtypes of missiles defined in the Sprite Set (*sam* and *bomb*) and it is crucial for the avatar to be able to shoot again (as *sam* sprites are Singleton).

The rest of the interactions in this game regulate when a sprite should be destroyed. Bases (structural defenses) are destroyed when colliding with a *bomb*, a *sam* and an *alien*. In the first two cases, the other sprite is also destroyed (via the effect KillBoth). The *avatar* is destroyed when colliding with an *alien* or a *bomb* (the enemies' missile). Finally, aliens are killed when hit by a *sam* (player's missile).

An important duty of the interaction set is to define the score system, which in turn can be used as reward for learning algorithms. Each interaction can define a parameter scoreChange which establishes how many points are awarded every time the interaction is triggered. Therefore, in *Aliens*, +1 point is given when the player destroys a *base* (line 22), +2 points when destroying an *alien* (line 26) and a point is lost when the avatar is destroyed (line 25).

The **Termination Set** (lines 28–30) describes the ways in which the game can end. All instructions must define the parameter win, which determines if the game is won or lost from the perspective of the player. The first condition (line 29) is of type SpriteCounter, which checks that the count of one particular type of sprites reaches the limit. In this case, this instruction determines that, when the player is killed, the game ends in a loss. The second condition (line 30), a MultiSpriteCounter, requires that both types of sprites (stype1 and stype2) have no instances in the game for the game to end. In this case, when there are no aliens in the game and the portal is gone (i.e., it has spawned all aliens already), the game is finished with a win for the player.

Finally, the **LevelMapping** set describes the sprites that correspond to the characters defined in the level file. All VGDL games can be played in many levels, which set the initial configuration of the sprites when the game starts. Figure 2.1 shows two examples of levels for *Aliens* available in the framework.

Each character in the level definition and mapping can define one or more sprites that will be created at that position. In the case of *Aliens*, all characters define a common sprite: *background*. This sprite, defined in the Sprite Set, has the only purpose of being drawn behind all sprites (so when, for instance, a *base* is destroyed, something is drawn instead). Note also that having two different types of portals allows for the creation of different levels with different difficulties. *portalFast* (character "2") spawns aliens at a higher rate (making the game more difficult) than *portalSlow* (resp. "1"). The two levels in Figure 2.1 use different portals (easier

```
 1   1................................          1   2................................
 2   0 0 0............................          2   0 0 0............................
 3   0 0 0............................          3   0 0 0............................
 4   ................................          4   ................................
 5   ................................          5   ......0000........0000......
 6   ................................          6   ......0..0........0..0......
 7   ................................          7   ................................
 8   ....000......000000.....000...          8   ................................
 9   ...00000....00000000...00000..          9   ..00000....00000000....00000..
10   ...0...0....00...00...00000..         10   ..0...0....00...00....00000..
11   ..................A...........         11   ..............A...........
```

Figure 2.1: Two VGDL levels for the game *Aliens*.

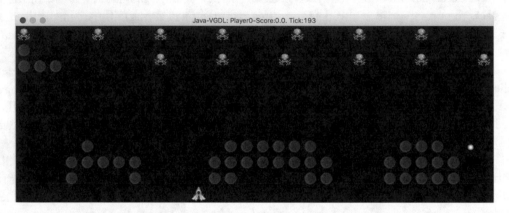

Figure 2.2: Screenshot of the game *Aliens* in GVGAI.

case for the level on the left, harder for the one on the right). Figure 2.2 shows a screenshot of the *Aliens* game at an intermediate stage.

VGDL is a rich language that allows the definition of multiple 2D games in which sprites can move actively (through behaviors specified in the ontology or the actions of the player), passively (subject to physics), and collide with other objects that trigger effects. These interactions are local between two colliding objects, which can lead to some sprites disappearing, spawning, transforming into different types or changing their properties. Non-local interactions can also be achieved, as teleportation events or effects that affect multiple sprites (all, or by type) are possible.

The VGDL underlying ontology permits descriptions to be concise. This ontology defines high-level building blocks for games, such as physics, movement dynamics and interactions. Given these tools, VGDL is able to describe a wide variety of arcade games, including versions of classic arcades such as *Space Invaders* (described here), *Sokoban*, *Seaquest*, *Pong*, *Pac-Man*, *Lunar Lander* or *Arkanoid*.

Some of these games, however, could not be defined in the original VGDL for GVGAI. The following two sections describe the introduction of real-world physics (Section 2.2.1), to include effects like gravity and inertia, and the ability to define 2-player games (Section 2.2.2).

2.2.1   CONTINUOUS PHYSICS

The type of games that the original VGDL could describe for GVGAI is limited to arcade games with discrete physics. All games before 2017 were based on a grid structure that discretizes the state of the game and the available actions. For instance, the orientation of the sprites described above could be either `Nil`, `Up`, `Down`, `Left` or `Right`, without providing a finer precision. While this is enough to create games like *Aliens*, it falls short for games such as Lunar Lander, which requires finer precision for orientation and movement.

Pérez Liébana et al. [2017] introduced in 2017 an update in VGDL that enhanced the ontology to widen the range of games that can be built. The main addition to the framework was the ability of establishing a *continuous* physics setting that would allow sprites to move in a more fine-tuned manner. Sprites can now be asked to adhere to the new physics type by setting a new parameter `physicstype` to `CONT`. Sprites that align to these physics acquire three extra parameters that can be defined: `mass`, `friction` and `gravity`. Although it is normal for gravity to be set equal to all sprites, it can also be set separately for them.

Movement of sprites is still active, like in the case of the traditional grid physics, but continuous physics do not limit movement to four directions. Instead, orientation can be set to 360°. Furthermore, speed is not defined proportional to cell but to screen pixels, allowing the sprite to move in any possible direction and with and speed.

Sprites are also subject to a passive movement, which occurs at every game frame. This passive movement alters the direction of the sprite, either by friction or by gravity. If the sprite is affected by friction, its speed will be multiplied by $1 - f(s)$ (where $f(s)$ is the sprite's friction). In case it responds to gravity, a vector $\vec{g}$ that points downwards and has a magnitude equal to the sprite's mass multiplied by the sprite's gravity is added to the sprite's current velocity vector.

Figure 2.3 shows an example of a game built with continuous physics, *Arkanoid*. In this game, the player controls the bat in the bottom of the screen. It must bounce a ball to destroy all the bricks in the level in order to win. The player loses a life when the ball falls through the bottom end, losing the game when all lives have been lost.

Listing 2.2 shows a simplified version of the Sprite Set for this game. The avatar sprite, of type `FlakAvatar` (line 6), defines the types of physics to be continuous. `friction` and `inertia` are set to 0.2, which establish the dynamics of the bat movement. Other interesting parameters are the ones that define the number of lives the player has (`healthPoints` and `limitHealthPoints`) and the width of the bat (`wMult`).

The *ball* (of type `Missile`) also adheres to this type of physics. For this game, no gravity is needed (the ball in *Arkanoid* does not really obeys the law of physics), but `speed` is defined at 20 speeds per second.

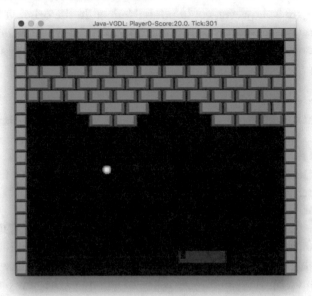

Figure 2.3: Screenshot of the game *Arkanoid* in GVGAI.

```
1    SpriteSet
2        background > Immovable img=oryx/space1 hidden=True
3
4        avatar > FlakAvatar stype=ball physicstype=CONT wMult=4
5            friction=0.2 mass=0.2 img=oryx/floor3 healthPoints=3
6            limitHealthPoints=3
7
8        ball > Missile orientation=UP speed=20 physicstype=CONT
9        ballLost > Passive invisible=True hidden=True
10
11       brick > Passive img=newset/blockG wMult=2
12       block > Passive img=newset/block2 wMult=2
```

Listing 2.2: VGDL Sprite Set block of the game *Arkanoid*.

2.2.2 2-PLAYER GAMES

The original VGDL only allowed for single-player games to be created. Gaina, Pérez Liébana, and Lucas [2016] and Gaina et al. [2017a] introduced the possibility of defining 2-player simultaneous games in this language.

One of the first modifications to the language was the addition of a new parameter (no_players) to specify the number of players that the game is for. This is added to the first line of the game description, as Listing 2.3 shows.

```
1 BasicGame square_size=50 no_players=2
```

Listing 2.3: Specifying the number of players in VGDL.

```
1 avatarA npc > killSprite scoreChange=-1,0
2 avatarB npc > killSprite scoreChange=0,-1
3
4 SpriteCounter stype=avatarA limit=0 win=False,True
5 SpriteCounter stype=avatarB limit=0 win=True,False
```

Listing 2.4: Score and termination instructions for a 2-player VGDL game.

Additionally, the score interactions must specify the number of points each player receives when an event is triggered and the termination instructions have to determine he end state (win or lose) for each player. In both cases, this is separated by a comma, as Listing 2.4 shows.

An important aspect to note is that games can be cooperative or competitive, which is inferred from the interaction and termination sets that describe the game. Figure 2.4 shows two examples of 2-player games in GVGAI. The one on the left, *Minesweeper*, in which two players compete for finding their own bombs in boxes scattered across the level. In the one of the right, *Sokoban*, two players cooperate to push all crates into target holes.

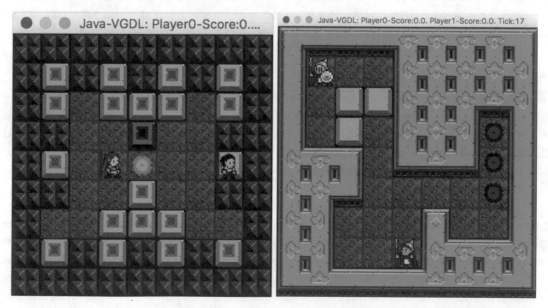

Figure 2.4: Examples of two 2-player VGDL games: *Minesweeper* (left) and *Sokoban* (right).

2.2.3   THE FUTURE OF VGDL

By July 2019, 140 single-player and 60 two-player VGDL games have been created, with 5 different levels per game. As can be seen in the rest of this book, VGDL has been a crucial part of GVGAI, as a rapid language to prototype and design new challenges. The integration of continuous physics and two-player capabilities have widened the variety of possible games that can be created.

However, VGDL is still limited. Despite its expressiveness, it is hard to make VGDL games that are truly fun for humans to play. Furthermore, creating complex games is also limited in the language, which in turn puts a limit in the type of challenges that can be posed for intelligent agents. Apart from interfacing GVGAI with non-VGDL games (see Chapter 8), further work in VGDL is envisioned.

One of the potential changes to VGDL is to adapt it for games in which the player doesn't necessarily need to be in control of an avatar (which is a requirement in the current version). For instance, games like Tetris or Candy Crush can't be created in the current version of VGDL without an important number of instructions and hacking. Another logical next step is to move toward the possibility of creating 3D games, as well as moving from 2-player to N-player challenges.

As will be seen later in Chapters 6 and 7, VGDL is used for content generation and game design. VGDL was not originally designed for these kind of tasks, and as such it is not trivial to adapt it for them. Advances toward a more modular language can ease the automatic design of rules, levels and games and facilitate the creation of new interesting games and challenges.

2.3   THE GENERAL VIDEO GAME AI FRAMEWORK

The original implementation of VGDL was parsed in `py-vgdl`, a Python interpreter written by Schaul [2013b] that permitted agents to play the original VGDL games. In order to provide a fast Forward Model for planning agents, py-vgdl was ported to Java for the first GVGAI framework and competition. This framework is able to load games and levels defined in VGDL and expose API for bots with access to an FM. Figure 2.5 shows a scheme of the GVGAI framework at work.

The Java-vgdl engine parses the game and level description to creates two components: a game object and a controller, which can be a interfaced by a bot or a human player. Then, the game starts and request an action from the controller at each game frame. The controller must return an action to execute in the real game in no more than 40 ms, or penalties apply. If the action is returned in between 40 and 50 ms, a null action (NIL) will be applied. If the action is obtained after more than 50 ms, the controller is disqualified. This rule is set up in order to preserve the real-time nature of the games played.

GVGAI bots must implement a class according to a given API. In single-player games, this class must inherit from the abstract class `AbstractPlayer` (`AbstractMultiPlayer` for two-player games). The API for single-player games controllers is included in Listing 2.5.

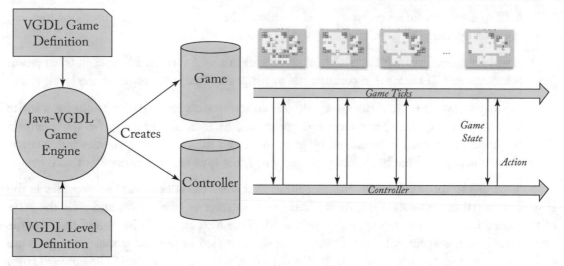

Figure 2.5: The GVGAI framework.

```
1    public  <ClassName >( StateObservation ,  ElapsedCpuTimer );
2
3    Types.ACTIONS  act ( StateObservation ,  ElapsedCpuTimer );
4
5    void  result ( StateObservation ,  ElapsedCpuTimer );
```

Listing 2.5: API for single-player games.

The function act is called from the game at every game frame and must return a
Types.ACTIONS value at every 40 ms. Although each game defines a different set of available actions, all games choose a subset of all possible moves: RIGHT, LEFT, UP, DOWN, USE and
NIL. The first four are movement actions, USE is employed as a wild card for non-movement
actions that the avatar type may define (for instance, shooting) and NIL is used to return a void
action. The second function, result, is optional and is called only when the game is over.

This function act, however, *must* be implemented in the bot. Both functions receive two
parameters: a StateObservation and a timer. The timer can be queried to control the time left
until the action needs to be returned. The state observation provides information about the state:

- Current time step, score, victory state (*win*, *loss* or *ongoing*), and health points of the player.

- The list of available actions for the player.

- A list of observations. Each one of the sprites of the game is assign a unique integer type
  and a category attending to its behavior (static, non-static, non-player characters, col-
  lectibles and portals). The observation of a sprite, unless it's a hidden one, contains infor-

mation about its category, type id and position. These observations are provided in a list, grouped by category (first) and sprite type (second).

- An observation grid with all visible observations in a grid format. Here, each observation is located in a 2D array that corresponds to the positions of the sprites in the level.

- History of avatar events. An *avatar event* is an interaction between the avatar (or a sprite produced by the avatar) and any other sprite in the game. The history of avatar events is provided in a sorted list, ordered by game step, and each instance provides information about the place where the interaction happened, the type and the category of each event.

One of the most important functionalities that the `StateObservation` provides is the FM. The function `advance` from `StateObservation` receives one action and rolls the state of the game forward one step applying such action. The next state will be one of the possible next states that the agent will find if that action were applied in the real game. The games are stochastic in nature, so unless the game is fully deterministic, the next state will be sampled from the possible next states that could exist. Additionally, `StateObservation` provides the function `copy`, which allows to make an exact copy of the current state. These two functions are essential to be able to run statistical forward planning (SFP) algorithms like MCTS or RHEA.

Apart from the function `act`, the controller must also implement a constructor that receives the same parameters. This constructor builds the agent and is called just before the game starts. This function must finish in 1 s and can be used to analyze the game and initialize the agent. It's important to notice that the agent never has access to the VGDL description of the game—it must guess the rules, dynamics and ways to win using the FM.

2.3.1 API FOR 2-PLAYER GAMES

Bots created for the 2-player games need to adhere to a different API. These are the main differences.

- The class that implements the agent must inherit from the base class `AbstractMultiPlayer`.

- The same functions as in the single-player case are available, although in this case the state observation is received as a `StateObservationMulti` object.

- Each player receives a player ID in the form of an integer, which is supplied as a third parameter in the constructor. This player ID can be 0 or 1 and it is assigned by the game engine. Listing 2.6 shows the API for a 2-player games agent.

- Score, victory status, health points and list of available actions are still available through the `StateObservationMulti` object. However, the respective methods now receive a parameter: the player ID from the player whose information is requested. Note that the agents may have a different set of actions available.

```
1    public <ClassName>(StateObservationMulti, ElapsedCpuTimer, int);
2
3    Types.ACTIONS act(StateObservationMulti, ElapsedCpuTimer);
4
5    void result(StateObservationMulti, ElapsedCpuTimer);
```

Listing 2.6: API for single-player games.

- Access to an FM is available as in the single-player case, but the function advance now receives an array of actions for all players in the game. The index in this array corresponds to the player ID and the actions are executed simultaneously to roll the state forward.

Agents in 2-player games may have different objectives and ways to win the game. The API does not provide information about the cooperative or competitive nature of the game, leaving the discovery of this to the agents that play the game.

2.3.2   GVGAI AS A MULTI-TRACK CHALLENGE

The original use of the GVGAI framework was for playing single-player **planning**, later expanded to also cover 2-player games. We refer to this setting (or track) as *planning* because the presence of the FM permits the implementation of statistical forward planning methods such as MCTS and RHEA. The agents do not require previous training in the games, but the FM allows the bots to analyze search trees and sequences of actions with the correct model of the game. Chapter 3 shows simple and advanced methods that can be used for the planning challenge in GVGAI.

A different task is to learn to play games episodically without access to an FM. The **learning** track was later added to the framework to tackle this type of learning problem. For this task, agents can be also written in Python apart from Java. This was included to accommodate for popular machine learning (ML) libraries developed in Python. With the exception of the FM, the same information about the game state is supplied to the agent, but in this case it is provided in a JavaScript Object Notation (JSON) format. Additionally, the game screen can also be observed, for learning tasks via pixels only. Torrado et al. [2018] interfaced the GVGAI framework to the OpenAI Gym environment for a further reach within the ML community. Chapter 5 dives deeper in this setting of GVGAI.

The GVGAI framework has primarily be used for agents that play games, but games as an end in themselves can also be approached. In particular, the framework was enhanced with interfaces for automatic **level** and **rule** generation. These are pocedural content generation (PCG) tracks where the objective is to build generators that can generate levels for any game received and (respectively) create rules to form a playable game from any level and sprite set given. In both cases, access to the FM is granted so the generators can use planning agents to evaluate the generated content. Chapter 6 describes these interfaces in more detail.

GVGAI thus offers a challenge for multiple areas of Game AI, each one of them covered by one of the settings described above. The planning tracks offer the possibility of using model-based methods such as MCTS and RHEA and, in the particular case of 2-player games, proposes research in problems like general opponent modeling and interaction with another agent. The learning track promotes the investigation in general model-free techniques and similar approaches such neuro-evolution. Last but not least, the content generation settings put the focus in algorithms typically used for PCG tasks, such as solvers (Satisfiability Problems and Answer Set Programming), search-based methods (Evolutionary Algorithms and Planning), grammar-based approaches, cellular automata, noise and fractals.

2.4   THE GVGAI COMPETITION

Each one of the settings described in the previous sections has been proposed as a competition track for the game AI community. Table 2.1 summarizes all editions of the competition until 2018.

Unless it's specified differently, all tracks follow a similar structure: games are organized into sets of 10 (plus 5 levels per game). N sets[1] are made public and available within the framework code.[2] Participants can therefore use those games to train their agents.

One extra set is used for *validation*. Its games are unknown and not provided to the participants, although they can submit their agents to the competition server[3] to be evaluated in them (as well as in the training sets). The server provides rankings of the agents based on training and validation sets, keeping the validation games anonymized. Finally, a hidden and secret *test* set is used to evaluate the agents and form the final rankings of the competition. Participants can't access these games nor evaluate their agents on them until submissions are closed. Once the competition is over, the games and levels of the validation set are released and added to the framework, the test set will be used as validation for the next edition of the competition and 10 new games will be created to form the next test set.

Competition rankings, in a set of games, are computed as follows: first, all the entries play each game in all the available levels (once for training and validation sets, 5 in the test set). All games end after 2000 frames with a loss for the player if the natural game end conditions are not triggered before. A game ranking is established by sorting all the entries first by average of victories, using average score and game duration as tie breakers, in this order. All indicators but the latter are meant to be maximized. For the time-steps count, t are counted for games that end in a victory and $2000 - t$ for those ended in a loss; this rewards games being won quickly and losses reached late.

For each game ranking, points are awarded to the agents following an F1 system: from first to tenth position, $25, 18, 15, 12, 10, 8, 6, 4, 2$ and 1 points are given, respectively. From the

---

[1]N depends on the competition. The collection of GVGAI games grows by 10 (1 set) for each successive edition.
[2]Competition Framework: https://github.com/GAIGResearch/GVGAI.
[3]Competition website: http://www.gvgai.net/.

Table 2.1: All editions of the GVGAI competition, including winners, approach, achievements and number of entries submitted (sample agents excluded). All editions ran in a different set of games and some competition winners did not participate in all editions.

Single-Player Planning				
Contest	**Winner**	**Approach**	**% Victories**	**Entries**
CIG 2014	adrienctx	Open Loop Expectimax Tree Search (OLETS)	51.2	14
GECCO 2015	YOLOBOT	MCTS, BFS, Sprite Targeting Heuristic	63.8	47
CIG 2015	Return42	Genetic Algorithms, Random Walks, A*	35.8	48
CEEC 2015	YBCriber	Iterative Widening, Danger Avoidance	39.2	50
GECCO 2016	MaastCTS2	Enhanced MCTS	43.6	19
CIG 2016	YOLOBOT	(see above)	41.6	24
GECCO 2017	YOLOBOT	(see above)	42.4	25
WCCI 2018	asd592	Genetic Algorithms, BFS	39.6	10
Two-Player Planning				
Contest	**Winner**	**Approach**	**Points**	**Entries**
WCCI 2016	ToVo2	SARSA-UCT with TD-backups	178	8
CIG 2016	adrienctx	OLETS	142	8
CEC 2017	ToVo2	(see above)	161	12
FDG 2018	adrienctx	(see above)	169	14
Level Generation				
Contest	**Winner**	**Approach**	**% Votes**	**Entries**
IJCAI 2016	easablade	Cellular Automata	36	4
CIG 2017	(suspended)	-	-	1
GECCO 2018	architect	Constructive and Evolutionary Generator	64.74	6
Rule Generation				
Contest	**Winner**	**Approach**	**% Votes**	**Entries**
CIG 2017	(suspended)	-	-	0
GECCO 2018	(suspended)	-	-	0

tenth position down, no points are awarded. The final rankings are computed by adding up all the points achieved in each game of the set. Final ranking ties are broken by counting the number of first positions in the game rankings. If the tie persists, the number of second positions achieved is considered, and so on until the tie is broken.

In the 2-player planning case, the final rankings are computed by playing a round-robin tournament among all submitted agents in the final set, with as many iterations as time allows. In this case, all levels are played twice, swapping positions of the agents to account for possible unbalance in the games. Training and validation sets are run differently, though, as it wouldn't be possible to provide rapid feedback if a whole round-robin needs to be run every time a new submission is received. In this case, a Glicko-2 score system [Glickman, 2012] is established for selecting the next two agents that must play next. In these sets, Glicko-2 scores become the first indicator to form the rankings. The test set also provides Glicko-2 scores but only as an extra indicator: final rankings are computed as in the planing case (victories, score and game duration).

The learning track has been run in two different editions (2017 and 2018). For the 2017 case, controllers were run in two phases: learning and validation. In the learning stage, each entry has a limited time available (5 min) for training in the first three levels of each game. In the validation step, controllers play 10 times the other two levels of each game.

The framework used in the 2018 competition[4] interfaced with OpenAI Gym and the competition was run with some relaxed constraints. The set used for the competition only had three games and all were made public. Two of their levels are provided for training to the participants, while the other three are secret and used to obtain the final results. Also, each agent can count on 100 ms for decision time (instead of the traditional 40 ms). For full details on the learning track settings, the reader is referred to Chapter 5.

Finally, the winner of the PCG tracks is decided by human subjects who evaluate the content generated by the entries. In both settings, the judges are presented with pairs of generated content (levels or games) and asked which one (or both, or neither) is liked the most. The winner is selected based on the generator with more votes. For more details on the content generation tracks, please read Chapter 6.

2.5   EXERCISES

The GVGAI Framework is available in a Github repository.[5] Use the release 2.3[6] in order to run the same version presented in this chapter.[7]

---

[4]GVGAI-Gym Learning track framework: https://github.com/rubenrtorrado/GVGAI_GYM.
[5]https://github.com/GAIGResearch/GVGAI
[6]https://github.com/GAIGResearch/GVGAI/releases/tag/2.3
[7]These exercises are also available at this book's website: https://gaigresearch.github.io/gvgaibook/.

2.5.1   RUNNING GVGAI

One of the first activities that can be done with the GVGAI engine is to play the games. You can use the keyboard to play them or run one of the sample agents.

Playing as Human

Run `tracks.singlePlayer.Test`. It will automatically start the game Aliens giving the controls to a human player. You can control the ship with the arrow keys for movement and use SPACE to shoot.

Switching Games

You can try any other single-player game from running this file. To do so, simply open the file `examples/all_games_sp.csv` and take a look at the different games and annotate their index (number on the right of each row). Then, in the Test.java file, change the value of the variable `gameIdx` (line 37) to reflect this index, and run the program.

Playing as a Bot

To run any game within the collection using one of the sample planning agents, comment line 49 (which runs the game for human players) and uncomment line 52. The fourth parameter of the function called in this line is a string that represents the full path to the agent to run (by default is the Rolling Horizon Evolutionary Algorithm agent). Lines 18–26 define the strings for the sample agents included in the framework. Examples are the Monte Carlo tree search agent and OLETS, winner of several editions of the GVGAI competition.

2.5.2   VGDL

VGDL games are in the `examples` folder. There is a sub-folder for each game category: `gridphysics` and `contphysics` contain single-player games with traditional and real-world physics, `2player` has 2-player games and `gameDesign` contain games parameterized (see Chapter 7).

Editing a Game

Take one of the games in `examples/gridphysics` and open the VGDL description (file with the game name followed by '.txt'). Study the different parts of the VGDL description and try changing several values to see the effect they have when playing the game. Try also adding new rules (in the `InteractionSet`) or new sprites to modify the game to your liking. Maybe you can create a new game, variant of one of the games in the collection.

Editing a Level

You can also take one of the level description files (for instance, *aliens_lvl0.txt*) and change the layout of the level, or the initial position of the sprites. You can try taking one of the existing games and creating new levels for it of different difficulty levels.

2.5.3   SUBMIT TO THE GVGAI COMPETITION SERVER

In order to submit a bot to the GVGAI competition, you need to follow these steps.

- Create an account at the site, `http://www.gvgai.net/`.[8]

- Log in and go to *Submit → Submit to Single Player Planning*.

- Your controller should have file called Agent.java and included in a package with the same name as your username, in a .zip file. Full instructions can be found on the submission website.[9]

- Fill the submission form with the requested information. Then, select a game set to submit to. Your code will be compiled and executed on the server in the selected set(s) and you can follow the status of your submission on your profile homepage.

- Once your controller has finished playing (if there were no compilation or execution errors) you can check the rankings page to compare your entry to others.

- If you wish to take part in one of the competitions, you can test this procedure first using one of the sample controllers distributed with the framework.

---

[8]There is absolutely no spamming associated with it.
[9]`http://www.gvgai.net/submit_gvg.php`

CHAPTER 3

Planning in GVGAI

Diego Pérez Liébana and Raluca D. Gaina

3.1 INTRODUCTION

This chapter focuses on AI methods used to tackle the planning problems generated by GVGAI. Planning problems refer to creating plans of action so as to solve a given problem; for example, figuring out how to first pick up a key and use it to open the exit door of a level while avoiding enemies, in an adventure game like Zelda. For this purpose, a model of the environment is available to simulate possible future states, when given a current state and an action to be taken by the player; this model will be referred to as an FM from now on. Not all modern and complex games may have an FM available, an issue addressed in the Learning track of the framework; see Chapter 5. However, using the game engine to not only run the game, but also allow AI players to simulate states internally, is possible in many games and inherent to human intuition and processing of different scenarios. Creating a plan, imagining what may happen if that plan is executed and figuring out the best course of action is one method for human decision making which is widely applicable to a range of problems. Therefore, landing this ability to AI to reason in interesting scenarios, such as those depicted by GVGAI games, would take the technology one step closer to general intelligence.

In this chapter we explore several methods which apply and extend on this theory. MCTS and RHEA are the foundations on which most planning AI agents are built, featuring two different approaches: the first builds a game tree out of possible game actions and game states, from which it extracts statistics to decide what to do next in any given state; the second creates entire plans and uses evolutionary computation to combine and modify them so as to end up with the best option to execute in the game. These basic methods can be extended in several ways: knowledge learned about the environment can be imbued into the algorithm to better inform action plan evaluation, as well as other heuristics which allow the AI to make better than random decisions in the absence of extrinsic reward signals from the game. Additionally, GVGAI problems can be seen as multi-objective optimization problems: the scoring systems in the game may be deceptive and the AI would need to not only focus on gaining score (as in traditional arcade games), but also on solving the problem and winning the game, possibly even considering time constraints. Furthermore, the extensive research in Evolutionary Algorithms can be applied to RHEA methods for better results, for example by seeding the initial population with better than random action plans, by keeping a statistical tree similar to the approach taken by MCTS, by using statistics for modifying action plans, by keeping action plans in-between game ticks instead of starting from scratch every time or by dynamically modifying the action plan length depending on features of the current game state (i.e., exploring further ahead rewards vs. gathering enough statistics about close rewards to make informed decisions).

All of these enhancements are described in detail in the following sections of this chapter. In all experiments we compare the enhanced algorithms performance against the vanilla version as well as against other methods (i.e., RHEA vs. MCTS) to explore whether the modifications proposed improve upon the basic algorithm, as well as whether the new algorithm is able to compete with the state of the art and obtain better performance. The methods are tested on a subset of GVGAI games selected for the particular experiments, with the aim of either tackling specific game types or obtaining a diverse set of games proposing different types of challenges. All agents have 40 ms of decision time in the competition setting, and the works described here have normally respected this budget. In some cases, due to testing performed in multiple machines, an architecture-independent approach has been followed to avoid undesired bias in the experiments: instead of using time as budget, a maximum number of calls to the advance function of the FM is established.

This chapter is divided into four sections. Section 3.2 describes MCTS, followed by two sections showing some interesting variants for GVGAI (Knowledge-based and Multi-objective MCTS in Sections 3.3 and 3.4, respectively). Finally, Section 3.5 explores the use of RHEA and several enhancements.

3.2   MONTE CARLO TREE SEARCH

MCTS is a tree search algorithm initially proposed in 2006 [Chaslot et al., 2006, Coulom, 2006, Kocsis and Szepesvári, 2006]. Originally, it was applied to board games and is closely associated with Computer Go, where it led to a breakthrough in performance [Lee et al., 2009] before the irruption of deep learning and AlphaGo. MCTS is especially will suited for large branching factor games like Go, and this lead to MCTS being the first algorithm able to reach professional level play in the 9×9 board size version [Gelly and Silver, 2011]. MCTS rapidly became popular due to its significant success in this game, where traditional approaches had been failing to outplay experienced human players.

MCTS has since then been applied to a wide range of games, including games of hidden information, single-player games and real-time games. MCTS has also been used extensively in general game playing (GGP) [Finnsson, 2012] with very good results. An extensive survey that includes a description of MCTS, a complete collection of MCTS variants, and multiple applications of this algorithm to games and other domains, can be found at Browne et al. [2012].

The algorithm builds a search tree that grows in an asymmetric manner by adding a single node at a time, estimating its game-theoretic value by using self-play from the state of the node to the end of the game. Each node in the tree keeps certain statistics that indicate the empirical average ($Q(s, a)$) of the rewards obtained when choosing action a at state s, how often a move a is played from a given state s ($N(s, a)$) and how many times a state s has been visited ($N(s)$). The algorithm builds the tree in successive iterations by simulating actions in the game, making move choices based on these statistics.

Each iteration of MCTS is composed of several steps [Gelly et al., 2006]: *Tree selection, Expansion, Simulation* and *Backpropagation*. Figure 3.1 depicts these four steps in the algorithm.

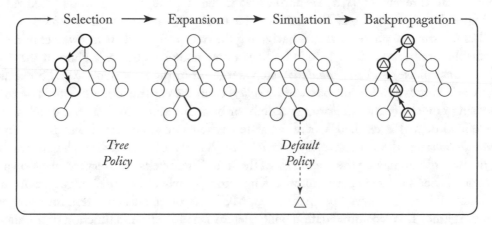

Figure 3.1: MCTS algorithm steps [Browne et al., 2012]: Selection, Expansion, Simulation and Backpropagation are executed iteratively until the allocated budget expires (time, iterations, or uses of the Forward Model).

At the start, the tree is composed only of the root node, which is a representation of the current state of the game. During the *selection* step, the tree is navigated from the root until a maximum depth or the end of the game has been reached. In this stage, actions are selected using a *multi-armed bandit* policy (tree policy) and applying it to the FM.

Multi-armed bandits is a known problem that features a slot machine with multiple arms. When an arm is pulled, a reward r is received drawn from an unknown probability distribution. The objective is to minimize the regret (or maximize the cumulative reward) when pulling from the different arms in sequence. In this scenario, regret is defined as the opportunity loss when choosing a sub-optimal arm. Good policies in this problem select actions by balancing the exploration of available arms and the exploitation of those that provided better rewards in the past. Auer, Cesa-Bianchi, and Fischer [2002] proposed the *Upper Confidence Bound* (UCB1, see Equation (3.1)) policy for arm bandit selection, and Kocsis and Szepesvári applied it later for tree search (UCT: Upper Confidence Bound for trees) [Kocsis and Szepesvári, 2006]:

$$a^* = \operatorname*{argmax}_{a \in A(s)} \left\{ Q(s, a) + C \sqrt{\frac{\ln N(s)}{N(s, a)}} \right\}. \tag{3.1}$$

The objective is to find an action a that maximizes the value given by the UCB1 equation. $Q(s, a)$ is the *exploitation* term, while the second term (weighted by the constant C) is the *exploration* term. The exploration term relates to how many times each action a has been selected from the given state s, and the amount of selections taken from the current state ($N(s, a)$ and $N(s)$, respectively). The value of C balances between exploration and exploitation. If the balancing weight $C = 0$, UCB1 behaves greedily following the action with the highest average outcome so far. If rewards $Q(s, a)$ are normalized in the range $[0, 1]$, a commonly used value for C in single-player games is $\sqrt{2}$. The optimal value of C may vary from game to game.

The *tree selection* phase continues navigating the tree until a node with fewer children than the available number of actions is found. Then, a new node is added as a child of the current one (*expansion* phase) and the *simulation* step starts. From the new node, MCTS executes a Monte Carlo simulation (or roll-out; *default policy*) from the expanded node. This is performed by choosing random (either uniformly random, or biased) actions until an end-game state (or a pre-defined depth) is reached, where the state of the game is evaluated. Finally, during the *backpropagation* step, the statistics $N(s)$, $N(s, a)$, and $Q(s, a)$ are updated for each node traversed, using the reward obtained in the evaluation of the state. These steps are executed in a loop until a termination criteria is met (such as number of iterations, or when the time budget is consumed).

Once all iterations have been performed, MCTS recommends an action for the agent to take in the game. This recommendation policy determines an action in function of the statistics stored in the root node. For instance, it could return the action chosen more often (a with the highest $N(s, a)$), the one that provides a highest average reward ($\mathrm{argmax}_{a \in A(s)} \; Q(s, a)$), or simply to apply Equation (3.1) at the root node. The pseudocode of MCTS is shown in Algorithm 3.1.

MCTS is considered to be an *anytime* algorithm, as it is able to provide a valid next move to choose at any moment in time. This is true independently from how many iterations the algorithm is able to make (although, in general, more iterations usually produce better results). This differs from other algorithms (such as A* in single-player games, and standard Min-Max for two-player games) that normally provide the next play only after they have finished. This makes MCTS a suitable candidate for real-time domains, where the decision time budget is limited, affecting the number of iterations that can be performed.

The GVGAI Framework provides a simple implementation of MCTS (also referred to here as *sampleMCTS* or *vanilla MCTS*). In the rest of this book, we refer to basic methods as "vanilla," i.e., methods containing the bare minimum steps for the algorithm, without any enhancements or tailoring for efficiency, performance or specific cases. In this MCTS implementation, C takes a value of $\sqrt{2}$ and rollout depth is set to 10 actions from the root node. Each state reached at the end of the *simulation* phase is evaluated using the score of the game at that point, normalized between the minimum and maximum scores ever seen during the play-outs. If the state is terminal, the reward assigned is a large positive (if the game is won) or negative (in case is lost) number.

Algorithm 3.1 General MCTS approach [Browne et al., 2012].

Input: Current Game State s_0
Output: Action to execute by the AI agent this turn.

1: **procedure** MCTS_Search(s_0)
2:     create root node v_0 with state s_0
3:     **while** within computational budget **do**
4:         v_l = TreePolicy(v_0)
5:         Δ = DefaultPolicy($s(v_l)$) {$s(v_l)$: State in node v_l, Δ: reward for state s.}
6:         Backup(v_l, Δ)
7:     **end while**
8:     **return** a(BestChild(v_0))
9:
10: **procedure** TreePolicy(v)
11:     **while** v is nonterminal **do**
12:         **if**(v not fully expanded)
13:             **return** *Expand*(v)
14:         **else**
15:             $v \leftarrow UCB1(v)$ {Eq. 3.1.}
16:             $s \leftarrow f(v(s), a(v))$
17:     **end while**
18:     **return** v
19:
20: **procedure** Expand(v)
21:     choose $a \in$ untried actions from $A(s(v))$ {$A(s(v))$: Available actions from state $s(v)$.}
22:     add a new child v' to v
23:         with $s(v') = f(s(v), a)$ {$f(s(v), a)$: State reached from $s(v)$ after applying a.}
24:     **return** v'
25:
26: **procedure** DefaultPolicy(s)
27:     **while** s is non-terminal **do**
28:         choose $a \in A(s)$ uniformly at random
29:         $s \leftarrow f(s, a)$
30:     **end while**
31:     **return** reward for state s
32:
33: **procedure** Backup(v, Δ)
34:     **while** v is not null **do**
35:         $N(s(v)) \leftarrow N(s(v)) + 1$
36:         $N(s(v), a(v)) \leftarrow N(s(v), a(v)) + 1$ {$a(v)$: Last action applied from state $s(v)$.}
37:         $Q(s(v), a(v)) \leftarrow Q(s(v), a(v)) + \Delta$
38:         $v \leftarrow$ parent of v
39:     **end while**

The vanilla MCTS agent performs relatively well in GVGAI games,[1] but with obvious limitations given that there is no game-dependent knowledge embedded in the algorithm. The value function described above is based exclusively in score and game end state, concepts that are present in all games and thus general to be used in a GVGP method. The following two sections propose two modifications to the standard MCTS to deal with this problem. The first one, Knowledge-based Fast Evolutionary MCTS (Section 3.3), explores the modification of the method to learn from the environment while the game is being played. The second one, proposes a Multi-Objective version of MCTS (Section 3.4) for GVGP.

3.3 KNOWLEDGE-BASED FAST EVOLUTIONARY MCTS

3.3.1 FAST EVOLUTION IN MCTS

Knowledge-based fast evolutionary MCTS (KB Fast-Evo MCTS) is an adaptation for GVGAI of a previous work by Lucas, Samothrakis, and Pérez Liébana [2014], who proposed an MCTS approach that uses evolution to learn a rollout policy from the environment. Fast-Evo MCTS uses evolution to adjust a set of weights w to bias the Monte Carlo simulations. These weights are used to select an action at each step in combination with a fixed set of features extracted for the current game state.

Each rollout evaluates a single individual (set of weights), using the value of the state reached at the end as the fitness. The evolutionary algorithm used to evolve these weights is a $(1 + 1)$ evolution strategy (ES). The pseudocode of this algorithm can be seen in Algorithm 3.2.

Algorithm 3.2 Fast Evolutionary MCTS Algorithm, from Lucas, Samothrakis, and Pérez Liébana [2014], assuming one roll-out per fitness evaluation.

Input: v_0 root state.
Output: weight vector w, action a.

1: **while** within computational budget **do**
2:     $w = Evo.GetNext()$
3:     Initialize Statistics Object S
4:     $v_l = TreePolicy(v_0)$
5:     $\delta = DefaultPolicy(s(v_l), D(w))$
6:     $UpdateStats(S, \delta)$
7:     $Evo.SetFitness(w, S)$
8: **end while**
9: **return** $w = Evo.getBest()$, $a = recommend(v_0)$

---

[1]See actual results in upcoming sections.

The call in line 3 retrieves the next individual to evaluate (w), a and the fitness is set in line 8. The vector of weights w is used to bias the rollout (line 6). For each state found in the rollout, first a number of N features are extracted (mapping from state space S to feature space F). Given a set of available actions A, a weighted sum of feature values determines the relative strength of each action (a_i), as shown in Equation (3.2).

Given that actions are weighted per feature, all weights are stored in a matrix W, where each entry w_{ij} is the weighting of feature j for action i. A softmax function (see Equation (3.3)) is used to select an action for the Monte Carlo simulation:

$$a_i = \sum_{j=1}^{N} w_{ij} \times f_i;$$ (3.2)

$$P(a_i) = \frac{e^{-a_i}}{\sum_{j=1}^{A} e^{-a_j}}.$$ (3.3)

The features selected to bias this action selection are euclidean distances from the avatar to the closest NPC, resource, non-static object and portal. In GVGAI, it is not possible to determine a feature space *a priori* (as the same method should work for any, even unknown, game). In fact, given that these features depend on the existence of certain sprites that may appear or disappear mid-game, the number of N features does not only vary from game to game, but also varies from game step to game step. Therefore, the algorithm needs to be flexible to adapt the vector of weights to the number of features at each state.

3.3.2   LEARNING DOMAIN KNOWLEDGE

The next step in defining KB Fast-Evo MCTS is to add a system that can provide a stronger fitness function for the individuals being evolved. Given that this fitness is calculated at the end of a rollout, the objective is to define a state evaluation function with *knowledge items* learned dynamically when playing the game.

In order to build this function, we define *knowledge base* as the combination of two factors: *curiosity* plus *experience*. For this work, *curiosity* refers to the discovery of the consequences of colliding with other sprites, while *experience* weights those events that provided a score gain. In both cases, the events logged are those where the avatar, or a sprite produced by the avatar, collides with another object in the game for which a feature is extracted (NPC, resource, non-static object and portal). Each one of these *knowledge items* keeps the following statistics.

- Z_i: number of occurrences of the event i.

- $\overline{x_i}$: average of the score *change*, which is the difference between the game score before and after the event took place. As GVGAI games are quite dynamic, multiple events happen at the same game tick and it is not possible to certainly assess which event actually triggered the score change. Therefore, the larger the number of occurrences Z_i, the more precise $\overline{x_i}$ will be.

These two values are updated after every use of the FM in the Monte Carlo simulation. When the state at the end of the rollout is reached, the following values are computed.

- Score change ΔR: this is the difference in game score between the initial and final states of the rollout.

- *Curiosity:* Knowledge change $\Delta Z = \sum_{i=1}^{N} \Delta(K_i)$, which measures the change of all Z_i in the knowledge base, for each knowledge item i. $\Delta(K_i)$ is calculated as shown in Equation (3.4), where Z_{i0} is the value of Z_i at the beginning of the rollout and Z_{iF} is the value of Z_i at the end:

$$\Delta(K_i) = \begin{cases} Z_{iF} & : Z_{i0} = 0 \\ \frac{Z_{iF}}{Z_{i0}} - 1 & : \text{otherwise.} \end{cases} \tag{3.4}$$

ΔZ is higher when the rollouts produce more events. Events that have been rarely triggered before will provide higher values of ΔZ, favoring knowledge gathering in the simulations.

- *Experience:* $\Delta D = \sum_{i=1}^{N} \Delta(D_i)$. This is a measure of change in the distance to each sprite of type i from the beginning to the end of the rollout. Equation (3.5) defines the value of $\Delta(D_i)$, where D_{i0} is the distance to the closest sprite of type i at the beginning of the rollout, and D_{iF} is the same distance at the end of the rollout:

$$\Delta(D_i) = \begin{cases} 1 - \frac{D_{iF}}{D_{i0}} & : Z_{i0} = 0 \text{ OR} \\ & \quad D_{i0} > 0 \text{ and } \overline{x_i} > 0 \\ 0 & : \text{otherwise.} \end{cases} \tag{3.5}$$

Note that ΔD will be higher if the avatar reduced the distance to those sprites with a positive $\overline{x_i}$ (i.e., provided a bost in score in the past) during the rollout, apart from reducing the distance to unknown sprites.

Equation (3.6) describes the final value for the game state and fitness for the individual being evaluated. This value is ΔR, unless $\Delta R = 0$. When $\Delta R = 0$, none of the actions during the rollout changed the score of the game, so the value must reflect the curiosity and experience components. This is done via a linear combination with weights $\alpha = 0.66$ and $\beta = 0.33$:

$$Reward = \begin{cases} \Delta R & : \Delta R \neq 0 \\ \alpha \times \Delta Z + \beta \times \Delta D & : \text{otherwise.} \end{cases} \tag{3.6}$$

Therefore, the new value function gives priority to actions that produce a score gain. However, when no score gain is achieved, actions that provide more information to the knowledge base or get the avatar closer to sprites that provide score are awarded.

3.3.3   EXPERIMENTAL WORK

Four different configurations of KB Fast-Evo MCTS have been tested on the 10 games of the first set of GVGAI games. Each game has 5 levels and each level has been played 5 times, totalling 250 games played for each one of the following configurations.

- **Vanilla MCTS**: the sample MCTS implementation from the framework, at the beginning of this section.

- **Fast-Evo MCTS**: fast Evolutionary MCTS, as per Lucas, Samothrakis, and Pérez Liébana [2014] (adapted to use a dynamic number of features).

- **KB MCTS**: knowledge-based (KB) MCTS as explained in the previous section, but using uniformly random selection in the rollouts (i.e., no evolution biases the Monte Carlo simulations).

- **KB Fast-Evo MCTS**: version of MCTS that uses both the knowledge base and evolution to bias the rollouts.

Performance in GVGAI is typically measured in two ways: percentage of games won and average score achieved in them. It's worthwhile highlighting that the former measure is easier to use for comparisons than the later, as each game has a different scoring system with different bounds and profiles.

Table 3.1 shows the win percentages achieved by each one of the four algorithm in the games used for testing. If we observe the total average of victories, *KB Fast-Evo MCTS* leads the comparison with 49.2% of games won. All the other variants achieved rates between 30.8% and 33.2%, showing that the addition of both the knowledge base and the fast evolution of weights to bias rollouts provides a boost in performance. However, the addition of any of them separately does not impact the algorithm significantly.

Looking at individual games, it can be seen that, in most games, *KB Fast-Evo MCTS* outperforms *Vanilla MCTS* in victory rate. In some games, the improvement can be observed as a consequence of adding either a stronger evaluation function or the fast evolution component (*Boulderdash* or *Zelda*, respectively). Although the addition of both parts also drives the improvement in other games (such as *Missile Command* and *Chase*).

Table 3.2 shows the average scores achieved by the same methods in these games. As in the case of the victory rate, *KB Fast-Evo MCTS* also gets higher total average score than the other methods, with 13.5 ± 1.2 points vs. the range 9–11 achieved by the other algorithms.

As mentioned above, it is more relevant to compare scores on a game by game basis due to the different score systems employed per game. In this case, *KB Fast-Evo MCTS* still outperforms *Vanilla MCTS* in most games in scores.

It can also be observed in the results that *KB Fast-Evo MCTS* fails to provide good results in certain games. Games like *Sokoban* and *Frogs* show little or no improvement at all. We hypothesize that the reasons for this are varied. One of them is the use of Euclidean distances

Table 3.1: Percentage of victories obtained on each game, standard error between parenthesis. In bold, those results that are the best ones on each game. Each value corresponds to the average result obtained by playing that particular game 25 times.

Game	Vanila MCTS	Fast-Evo MCTS	KB MCTS	KB Fast-Evo MCTS
Aliens	100.0 (0.0)	100.0 (0.0)	100.0 (0.0)	100.0 (0.0)
Boulderdash	0.0 (0.0)	4.0 (3.9)	**28.0 (9.0)**	**16.0 (7.3)**
Butterflies	88.0 (6.5)	**96.0 (3.9)**	80.0 (8.0)	**100.0 (0.0)**
Chase	12.0 (6.5)	12.0 (6.5)	0.0 (0.0)	**92.0 (5.4)**
Frogs	24.0 (8.5)	16.0 (7.3)	8.0 (5.4)	20.0 (8.0)
Missile Command	20.0 (8.0)	20.0 (8.0)	20.0 (8.0)	**56.0 (9.9)**
Portals	12.0 (6.5)	**28.0 (9.0)**	16.0 (7.3)	**28.0 (9.0)**
Sokoban	0.0 (0.0)	0.0 (0.0)	**4.0 (3.9)**	**8.0 (5.4)**
Survive Zombies	44.0 (9.9)	36.0 (9.6)	52.0 (10.0)	44.0 (9.9)
Zelda	8.0 (5.4)	**20.0 (8.0)**	8.0 (5.4)	**28.0 (9.0)**
Overall	30.8 (2.6)	33.2 (2.7)	31.6 (2.6)	**49.2 (3.2)**

for feature extraction. Using distances calculated by a path-finding algorithm such as A* would be more accurate, but the impact in the real-time nature of the algorithm would be quite high. Additionally, path-finding requires the definition of a navigable space, which is hard to describe as a general concept for *any* game.

However, it is likely that the major struggle is the relevance of the features taken for some of the games. For instance, in *Sokoban*, the avatar must push boxes to win the game and the orientation in which the box is pushed (and where it is pushed from) is relevant. But the features used here do not capture this information, as all collision events triggered when colliding with a box are treated the same. In other cases, like in *Frogs*, a specific sequence of actions must be applied so the avatar crosses the road without being hit by a truck. However, these sequences are rare—in most cases they collide with trucks, which ends in a game loss. Therefore the algorithm rewards not getting closer to these trucks and staying safe without incentives to cross the road.

3.4 MULTI-OBJECTIVE MCTS FOR GVGAI

This section proposes another modification to the MCTS algorithm for real-time games in which state evaluations consider two objectives instead of one. In GVGAI, these objectives are to maximize game score (same objective as in the vanilla MTCS described in Section 3.2) and to provide an incentive to maximize exploration in the game level. First, we provide some back-

Table 3.2: Scores achieved on each game, standard error between parenthesis. In bold, those results that are the best ones on each game. Each value corresponds to the average result obtained by playing that particular game 25 times.

Game	Vanila MCTS	Fast-Evo MCTS	KB MCTS	KB Fast-Evo MCTS
Aliens	36.72 (0.9)	38.4 (0.8)	37.56 (1.0)	**54.92 (1.6)**
Boulderdash	9.96 (1.0)	12.16 (1.2)	**17.28 (1.7)**	**16.44 (1.8)**
Butterflies	27.84 (2.8)	31.36 (3.4)	31.04 (3.4)	28.96 (2.8)
Chase	4.04 (0.6)	4.8 (0.6)	3.56 (0.7)	**9.28 (0.5)**
Frogs	-0.88 (0.3)	-1.04 (0.2)	-1.2 (0.2)	-0.68 (0.2)
Missile Command	-1.44 (0.3)	-1.44 (0.3)	-1.28 (0.3)	**3.24 (1.3)**
Portals	0.12 (0.06)	**0.28 (0.09)**	0.16 (0.07)	**0.28 (0.09)**
Sokoban	0.16 (0.1)	0.32 (0.1)	**0.7 (0.2)**	**0.6 (0.1)**
Survive Zombies	13.28 (2.3)	14.32 (2.4)	18.56 (3.1)	21.36 (3.3)
Zelda	0.08 (0.3)	0.6 (0.3)	0.8 (0.3)	0.6 (0.3)
Overall	9.0 (0.9)	10.0 (1.0)	10.7 (1.0)	**13.5 (1.2)**

ground in Multi-objective Optimization, to then describe the Multi-Objective MCTS (MO-MCTS) approach and the experimental work carried out to test this new algorithm.

3.4.1   MULTI-OBJECTIVE OPTIMIZATION

An optimization problem is called multi-objective when two or more conflicting objective functions must be optimized simultaneously. A multi-objective optimization problem can be defined as:

$$optimize \quad \{f_1(\vec{x}), f_2(\vec{x}), \ldots, f_m(\vec{x})\} \tag{3.7}$$

subject to $\vec{x} \in \Omega$, with $m (\geq 2)$ conflicting objective functions $f_i : \Re^n \to \Re$. $\vec{x} = (x_1, x_2, \ldots, x_n)^T$ are *decision vectors* from the feasible region $\Omega \subset \Re^n$. $Z \subset \Re^m$ is the feasible objective region. The elements of this region are known as *objective vectors*, which consist of m objective values $\vec{f}(\vec{x}) = (f_1(\vec{x}), f_2(\vec{x}), \ldots, f_m(\vec{x}))$. Each solution $\vec{x}$ results in a set of m different values to be optimized.

One solution $\vec{x}$ is said to *dominate* another solution $\vec{y}$ if and only if:

1. $f_i(\vec{x})$ is not worse than $f_i(\vec{y})$, $\forall i = 1, 2, \ldots, m$; and

2. $f_j(\vec{x})$ is better than its analogous counterpart in $f_j(\vec{y})$ in at least one objective j.

If these two conditions are met, it is said that $\vec{x} \prec \vec{y}$ ($\vec{x}$ *dominates* $\vec{y}$). This condition determines a *partial ordering* between solutions in the objective space. In the case where it is not possible to state that $\vec{x} \prec \vec{y}$ or $\vec{y} \prec \vec{x}$, it is said that these solutions are *indifferent* to each other. Solutions indifferent to each other form part of the same *non-dominated set*. A non-dominated set P is said to be the *Pareto-set* if there is no other solution in the decision space that dominates any member of P. The objective vectors of P build a *Pareto-front*.

One of the most popular methods to measure the quality of a non-dominated set is the hypervolume indicator (HV). This indicator measures both the diversity and convergence of non-dominated solutions [Zitzler, 1999]. The HV of a Pareto front P ($HV(P)$) is defined as the volume of the objective space dominated by P. The higher the value of $HV(P)$ the better the front is, assuming all objectives are to be maximized.

3.4.2   MULTI-OBJECTIVE MCTS

MO-MCTS [Pérez Liébana et al., 2014] tackles the problem of selecting an action with a reduced time budget in an MO setting. The algorithm requires that a game state is evaluated according to m objectives, returning a vector $\vec{r}$. $\vec{r}$ is the reward vector to be used in the *backpropagation* step of MCTS through all the nodes visited in the last iteration. This vector updates an accumulated reward vector $\overline{R}$ and a local Pareto front approximation P at each node. Algorithm 3.3 describes how the update of the node statistics in MO-MCTS.

Algorithm 3.3 Node update in the backpropagation step of MO-MCTS [Pérez Liébana et al., 2014].

Input: node current tree node being updated
Input: $\overline{r}$ reward of the last state evaluation
Input: dominated if $\overline{r}$ is dominated by a front from the descendants of *node*

 1: **procedure** Update(*node*, $\overline{r}$, *dominated = false*)
 2:     *node.Visits = node.Visits* $+ 1$
 3:     *node.*$\overline{R}$ = *node.*$\overline{R}$ $+ \overline{r}$
 4:     **if** !*dominated* **then**
 5:         **if** *node.P* $\prec \overline{r}$ **then**
 6:             *dominated = true*
 7:         **else**
 8:             {P is the Pareto front approximation at each node}
 9:             *node.P = node.P* $\cup \overline{r}$
10:         **end if**
11:     **end if**
12:     *Update(node.parent,* $\overline{r}$, *dominated*)

The Pareto front P update (line 9 of Algorithm 3.3) in a node works as follows: in the case that $\bar{r}$ is *not* dominated by the front, $\bar{r}$ is added to P. This update considers that some of the current points in P may leave the set if $\bar{r}$ dominates any of them. If $\bar{r}$ is dominated by P, the front remains unchanged. Note that, if the latter is true, no further fronts will be changed in all remaining nodes until reaching the root.

Each node can have an estimation over the quality of the reachable states by keeping its local front P updated. The quality of the front is calculated as the hyper-volume $HV(P)$, which substitutes the exploitation term $Q(s, a)$ in the MO-UCB Equation (3.8):

$$a^* = \underset{a \in A(s)}{\operatorname{argmax}} \left\{ HV(P) + C \sqrt{\frac{\ln N(s)}{N(s, a)}} \right\}. \tag{3.8}$$

As the backpropagation step is followed until reaching the root node, it is straightforward to see that the root contains the best non-dominated front of the whole search. Therefore, the root note can also provide information to the recommendation policy to select an action to play in the game once the budget is over. In MO-MCTS, the root stores information about which action leads to which point in its own non-dominated front. Weights can then be defined to determine which point of the front P at the root is chosen and pick the action that leads to it.

Heuristics for GVGAI

This section proposes two heuristics for GVGAI methods, each one of them will be treated as objectives in the experimental testing for this approach.

Score (Objective O_1) This is the simple heuristic describe above that uses on the game score and the game end condition: the value of a state is the current score unless the game is over. In that case, a large integer is added if the game is won (10^6) or lost (-10^6).

Level Exploration (Objective O_2) This heuristic computes a value that rewards an agent that maximizes the number of grid cells visited in the level. Its implementation is based on *pheromone* trails and it was first proposed in Pérez Liébana et al. [2015b]. The technique works as follows: the mechanism simulates that the avatar expels pheromones at each game tick, which spread into nearby cells. Each cell holds a pheromone value $p_{ij} \in [0, 1]$, where i and j are coordinates in the level grid. p_{ij} is set to decay with time, and the change of pheromone p_{ij} between steps is given by Equation (3.9):

$$p_{i,j} = \rho_{df} \times \rho_\phi + (1 - \rho_{df}) \times \rho_{dc} \times p_{i,j}, \tag{3.9}$$

ρ_ϕ is computed as the sum of pheromone trail in all surrounding cells divided by the number of neighboring cells.[2] $\rho_{df} \in (0, 1)$ sets the value of pheromone diffusion and $\rho_{dc} \in (0, 1)$ establishes the decay rate of the pheromone value at each frame. Values that showed good results previously are $\rho_{df} = 0.4$ and $\rho_{dc} = 0.99$.

---

[2]An edge cell has fewer neighbors.

This algorithm produces high values of pheromone trail in the close proximity of the avatar, as well as in positions recently visited. Figure 3.2 shows an example of how much pheromone is added to a cell and its neighbors.

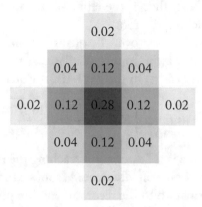

Figure 3.2: Pheromone diffusion [Pérez Liébana, Mostaghim, and Lucas, 2016]. The avatar is assumed to be located in the central position, where more pheromone is deposited. The neighboring cells receive less pheromone. These values are added to the pheromone previously deposited on each cell (values always kept $\in (0, 1)$).

This heuristic values higher positions where the pheromone value is small, which rewards agents that maximize the exploration of the level. Thus, the value of the heuristic is computed as $O_2 = 1 - p_{i,j}$, where i and j are the coordinates of the avatar in the grid. As in the previous case, a large positive or negative number is added to O_2 if the game is won or lost in a state—otherwise the agent would miss the opportunity of winning the game or, even worse, lose it in favor of exploring new positions.

3.4.3   EXPERIMENTAL WORK

Four different MCTS approaches are tested in this study. All these algorithms share the value of $C = \sqrt{2}$ for the UCB1/MO-UCB Equations (3.1) and (3.8), a limit of 50 iterations per game tick and a simulation depth of 10 moves from the root node. The experiments were conducted in the games of the first GVGAI set (10 games). Each approach described above played 100 times each one of the 5 levels, leading to 500 repetitions per game and agent. These approaches are listed next.

- Sample MCTS. The sample MCTS controller included in the GVGAI framework. The state value function is exactly O_1 in this case.

- Weighted Sum MCTS. A classical alternative to multi-objective approaches is to combine all the objectives into a linear combination. This implementation uses the default MCTS, but the state value function is determined as $O_1 \times \alpha + O_2 \times \beta$, with $\alpha = \beta = 0.5$.

- Mixed Strategy MCTS. The idea of this approach is to tackle the multi-objective problem as a mixed strategy [Myerson, 1991]. In this setting, each objective O_1, O_2 is managed by a different state evaluation function. The algorithm used is still sample MCTS but, at the beginning of the decision time, a higher-level policy determines which objective should be considered during that frame. In the experiments described here, the selection of one or the other is done uniformly at random.

- MO-MCTS. This approach implements the MO-MCTS algorithm described above, using O_1 and O_2 as the two objectives to optimize. The recommendation policy uses a linear combination $O_1 \times \alpha + O_2 \times \beta$ ($\alpha = \beta = 0.5$ as in the previous approaches) to evaluate each member of the Pareto front P owned by the root. The action that leads to the point with the highest weighed sum is picked to be played in the game.

Figure 3.3 shows the win rate of all agents in the games tested. It shows Sample MCTS and MO-MCTS as the strongest agents of the four, either of them achieving the highest percentage of victories on each game. MO-MCTS outperforms Sample MCTS in five games (*Aliens*, *Frogs*, *Missile Command*, *Portals* and *Zelda*).

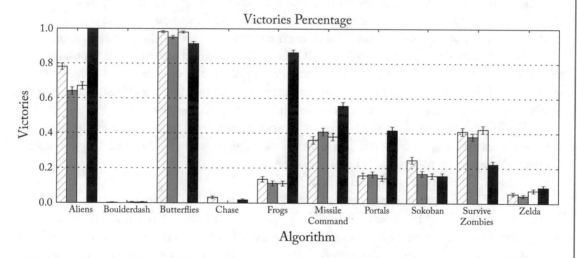

Figure 3.3: Percentage of victories (with std. error). Four approaches are compared per game. From left to right: Sample MCTS, Weighted Sum MCTS, Mixed Strategy MCTS, MO-MCTS. From Pérez Liébana, Mostaghim, and Lucas [2016].

MO-MCTS also achieves the highest win percentage if compared across all games to the other approaches. Table 3.3 shows this comparison, which indicates that Sample MCTS is the second best algorithm in this ranking.

Some interesting observations can be made when analyzing the results in a game per game basis. *Aliens*, for example, is a game that has traditionally been well played by Sample MCTS,

Table 3.3: Win rate achieved across all games in the first GVGAI game set

	Sample MCTS	Weighted Sum MCTS	Mixed Strategy MCTS	MO-MCTS
% Victories	32.24 (0.67)	29.80 (0.66)	30.51 (0.66)	**42.38 (0.70)**

achieving 100% of victories. This study uses 50 iterations per method, which approximately halves the budget that 40 ms per tick allows for this game in particular. With this reduction, the performance of Sample MCTS drops, while MO-MCTS achieves a 99,80% (0.20) win rate. Another impressive result is the high victory rate achieved by MO-MCTS in *Frogs*. This game has always posed many problems to MCTS approaches (see, for instance, Table 3.1 from Section 3.2), but MO-MCTS agent uses the exploration heuristic to find the goal of the level in 86.40% (1.53) of the games played.

In some cases, MO-MCTS achieves worse result than the other algorithms which seems to indicate that the excessive exploration can be a disadvantage. An example of this is *Survive Zombies*. In this game, a good strategy is to find a spot save from zombies and stay there. An excessive amount of exploration may lead to encounter more enemies and lose the game.

Not all games are favorable to MO-MCTS, however, and in some cases it is possible that the excessive exploration is actually a disadvantage. A good example could be *Survive Zombies*, where one of the best strategies is to locate a spot in the level safe from zombies. Exploring the level too much may lead to find more enemies and therefore to lose the game.

Figures 3.4 and 3.5 show the average of scores achieved by all approaches in the 10 games. Attending to this metric, MO-MCTS achieves the highest score in 7 out of 10 games, only beaten or matched by Sample MCTS on the other 3.

Table 3.4 shows the numerical data of these results.

MO-MCTS is significantly better in terms of scores than all approaches in seven games. It is worthwhile to highlight that MO-MCTS behaves clearly better (both in victories and in score) than the other multi-objective variants. These results seem to suggest that using multiple objectives to tackle GVGAI is promising, but the way these objectives are used is decisive to achieve good results. Combining them in a linear combination or using them in an alternative manner does not produce as good results as using the Pareto fronts in MO-MCTS.

Mixed Strategy MCTS does not achieve good results in this study, but a closer look at them having in mind the way this algorithm operates suggests some interesting insights. This controller spends 50% of its moves only focusing on new places to move to, without considering the score, and when this happens is determined at random. This is the only agent studied here that completely ignores the score in half of its moves. However, Mixed Strategy MCTS is significantly better than Weighted-Sum MCTS in two games (precisely in terms of score) and no worse in the other eight. This suggests that mixed strategies can work well in GVGP if

Table 3.4: Victory rate and score average (standard error). τ columns indicate significant dominance (Wilcoxon signed-rank test, p-value < 0.05) and bold font indicates dominance over the other three in victories or score. Games listed in the order used in Figure 3.3.

Game	Algorithm	Victories (%)	τ (Victory)	Scores	τ (Score)
G1	A: Sample MCTS	78.00 (1.85)	B , C	54.60 (0.39)	B
	B: Weighted Sum MCTS	64:00 (2:15)	Ø	53.91 (0.40)	Ø
	C: Mixed Strategy MCTS	67:00 (2:10)	Ø	54.44 (0.40)	Ø
	D: MO-MCTS	99.80 (0.20)	A , B , C	60.17 (0.50)	A , B , C
G2	A: Sample MCTS	0.25 (0.25)	Ø	6.83 (0.23)	Ø
	B: Weighted Sum MCTS	0.00 (0.00)	Ø	7.26 (0.26)	Ø
	C: Mixed Strategy MCTS	0.32 (0.32)	Ø	6.76 (0.25)	Ø
	D: MO-MCTS	0.40 (0.28)	Ø	7.53 (0.21)	Ø
G3	A: Sample MCTS	98.20 (0.59)	B , D	28.88 (0.62)	Ø
	B: Weighted Sum MCTS	95.00 (0.97)	D	28.84 (0.64)	Ø
	C: Mixed Strategy MCTS	98.00 (0.63)	B , D	30.02 (0.66)	Ø
	D: MO-MCTS	91.40 (1.25)	Ø	32.11 (0.72)	A , B , C
G4	A: Sample MCTS	3.20 (0.79)	B , C	2.47 (0.09)	B , C
	B: Weighted Sum MCTS	0.00 (0.00)	Ø	1.07 (0.06)	Ø
	C: Mixed Strategy MCTS	0.00 (0.00)	Ø	1.37 (0.06)	B
	D: MO-MCTS	1.80 (0.59)	B , C	2.49 (0.09)	B , C
G5	A: Sample MCTS	13.60 (1.53)	Ø	-1.36 (0.05)	Ø
	B: Weighted Sum MCTS	11.28 (1.47)	Ø	-1.41 (0.05)	Ø
	C: Mixed Strategy MCTS	11.20 (1.41)	Ø	-1.39 (0.05)	Ø
	D: MO-MCTS	86.40 (1.53)`	A , B , C	0.75 (0.03)	A , B , C
G6	A: Sample MCTS	36.00 (2.15)	Ø	0.68 (0.17)	Ø
	B: Weighted Sum MCTS	40.80 (2.20)	Ø	0.86 (0.19)	Ø
	C: Mixed Strategy MCTS	38.00 (2.17)	Ø	0.67 (0.17)	Ø
	D: MO-MCTS	55.60 (2.22)	A , B , C	3.47 (0.21)	A , B , C
G7	A: Sample MCTS	15.80 (1.63)	Ø	0.16 (0.02)	Ø
	B: Weighted Sum MCTS	16.40 (1.66)	Ø	0.16 (0.02)	Ø
	C: Mixed Strategy MCTS	14.20 (1.56)	Ø	0.14 (0.02)	Ø
	D: MO-MCTS	41.60 (2.20)	A , B , C	0.42 (0.02)	A , B , C
G8	**A: Sample MCTS**	24.60 (1.93)	B , C , D	1.28 (0.04)	B , C , D
	B: Weighted Sum MCTS	16.80 (1.67)	Ø	0.81 (0.04)	Ø
	C: Mixed Strategy MCTS	15.60 (1.62)	Ø	0.96 (0.03)	B
	D: MO-MCTS	15.60 (1.62)	Ø	1.09 (0.04)	B , C
G9	A: Sample MCTS	41.00 (2.20)	D	33.72 (1.31)	B
	B: Weighted Sum MCTS	38.00 (2.17)	D	29.92 (1.14)	Ø
	C: Mixed Strategy MCTS	42.20 (2.21)	D	32.54 (1.23)	Ø
	D: MO-MCTS	22.20 (1.86)	Ø	37.68 (1.21)	A , B , C
G10	A: Sample MCTS	5.40 (1.01)	Ø	2.26 (0.12)	Ø
	B: Weighted Sum MCTS	4.20 (0.90)	Ø	2.41 (0.13)	Ø
	C: Mixed Strategy MCTS	7.20 (1.16)	B	2.17 (0.12)	Ø
	D: MO-MCTS	9.00 (1.28)	A , B	3.69 (0.14)	A , B , C

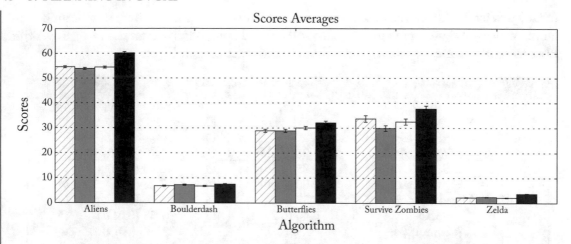

Figure 3.4: Average of scores (with std. error). Four approaches are compared per game. From left to right: Sample MCTS, Weighted Sum MCTS, Mixed Strategy MCTS, MO-MCTS. From Pérez Liébana, Mostaghim, and Lucas [2016].

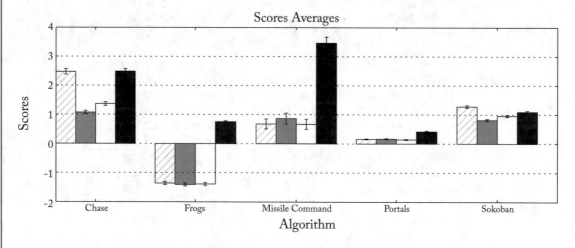

Figure 3.5: Average of scores (with std. error). Four approaches are compared per game. From left to right: Sample MCTS, Weighted Sum MCTS, Mixed Strategy MCTS, MO-MCTS. From Pérez Liébana, Mostaghim, and Lucas [2016].

objectives are chosen at the right time, encouraging further investigation on better (i.e., probably dynamic) balancing of the different objectives while playing.

3.5   ROLLING HORIZON EVOLUTIONARY ALGORITHMS

RHEA were first introduced by Pérez Liébana et al. [2013] as an alternative (and potentially better, more adaptive option) to MCTS for online planning in games. The concept they put forward was a novel usage of EAs for optimization of sequences (or plans) of actions in games. Therefore, the solution evolved by the EA is an action sequence of a specific length, which is executed for evaluation using simulations of a model of the game. This could be seen as similar to an MCTS rollout, where the final state reached after advancing through the actions in an individual is evaluated and gives its fitness value.

This technique has slowly become more popular in game AI research. Several authors applied RHEA to specific games, starting from the single-player real-time Physical Travelling Salesman Problem in 2013 [Pérez Liébana et al., 2013], to a two-player real-time Space Battle game in 2016 [Liu, Pérez Liébana, and Lucas, 2016], or single-player real-time games Asteroids and Planet Wars in 2018 [Lucas, Liu, and Pérez Liébana, 2018b]. Given that the algorithm performed well when adapted to multiple specific games, including the challenging multi-agent game Hero Academy [Justesen, Mahlmann, and Togelius, 2016], it seemed natural to test its strength in the GVGAI framework. Several works have been published on such applications in recent years, ranging from analysis of the vanilla algorithm, to enhancements or hybrids meant to boost performance. These modifications, which will be described in more detail in the following subsections, inspired the work to be extended to General Game Playing domains with moderate success by Santos, Bernardino, and Hauck [2018].

3.5.1   VANILLA RHEA

RHEA in its vanilla form follows several simple steps, at every game tick, as depicted in Figure 3.6.

1. **Population initialization**. All P individuals in the population are initialized as random action sequences of length L at the beginning of a game tick. For simplification and speed, individuals are represented by sequences of integers, where each gene can take a value between 0 and A, where A is the maximum number of available actions in the current game tick. Genes are always kept in this range through mutation, and are mapped back to game actions at the end of the evolution.

2. **Individual evaluation**. All individuals in the population are evaluated to assess their fitness. The actions in the sequence are executed, in turn, using a game model for simulations of possible future states given an action. The final state reached is evaluated with a heuristic H, and the value becomes the fitness of the individual.

3. **Order population** by fitness.

4. **Elitism**. The E best individuals are carried forward unchanged to the next generation.

5. **Individual selection**. Two parents are selected through tournament. [For $P > 1$].

6. **Offspring generation**. The parents are combined through uniform crossover to create a new individual (genes are randomly selected from the two parents to form a new individual). [For $P > 1$].

7. **Offspring mutation**. The offspring [or the only individual, if $P = 1$] is uniformly randomly mutated (genes are replaced with new random ones, with some probability M).

8. **Repeat steps 5–7** to create a new population of size P.

9. **Repeat steps 2–8** for N generations, or as long as the budget allows.

10. **Play first action of best individual** obtained at the end of the evolution process.

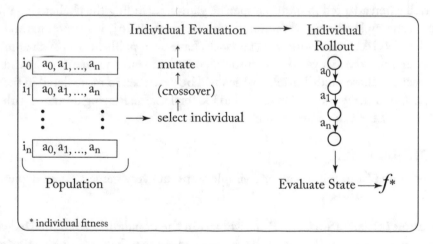

Figure 3.6: Rolling horizon evolutionary algorithm cycle.

In GVGAI, vanilla RHEA uses a simple heuristic function H to evaluate game states, represented by the current game score (which it aims to maximize), to which a large integer is added if the game state is final and RHEA won, or a large integer is subtracted if the game state is final and RHEA lost.

Parameter analysis   It is natural to notice several parameters in the steps described. Arguably the most important ones are P, the population size, and L, the length of the action sequences evolved. It is interesting to note that a popular approach in RHEA literature is to keep the population size to only 1 individual, turning the EA into a Random Mutation Hill Climber or $(1 + 1)$ EA: in this scenario, the one individual is mutated at every generation and it is kept (and the first is discarded) if its fitness is better than the first, or discarded otherwise. This leads to a focused fitness increase, with the risk of being stuck in local optima heavily relying on the chosen mutation operator.

Other parameters (such as elitism E or mutation rate M) can also influence the behavior of the algorithm, but research looked in depth at how P and L affect performance in a set of 20 GVGAI games, when all other parameters are fixed [Gaina et al., 2017b]. The values of these parameters were explored within a fixed budget of 480 calls to the forward model of the games, the average achieved by MCTS in the larger GVGAI game corpus in the 40 ms imposed by the framework for real-time decision making.

If both parameter values are increased so that only one generation can be created and evaluated, the algorithm becomes random search (RS)—that is, random action sequences are generated and the first action of the best sequence is played. No evolution takes place in this scenario.

The study carried out by Gaina et al. [2017b] found that, generally, the higher the L and P, the better—even in the extreme case where no evolution takes place. In the very limited budget, RHEA is unable to evolve better sequences than those randomly generated by RS, due to the big challenge of exploring quickly a large search space. In order for RHEA to be able to compete with RS and MCTS, it needs several enhancements that help it make better use of the budget and therefore search more efficiently (some will be detailed in the following section). The study does highlight that, given more budget, RHEA is able to outperform RS—therefore it is indeed simply a case of fast and efficient evolution being needed.

It is interesting, however, that high population sizes in RHEA lead, on average, to its ability to outperform MCTS nonetheless. Since most GVGAI competition entries (including past winners) are based on MCTS with several modifications, this is an indication that RHEA-based entries have the potential to be even better and definitely a viable alternative for general AI.

RHEA with very long sequences was also shown to perform better in Gaina, Lucas, and Pérez Liébana [2019], when budget is increased proportionally as well—although going beyond 150 appears to be detrimental especially in dense reward games.

Although on average more and longer individuals seem better, this does differ on a game by game basis: for example, in very dense reward games it may be more beneficial to have shorter individuals, since the reward landscape varies enough in the short-term for the agent to be able to tell which are good actions and which are not (although this could affect its performance in deceptive games in which short-term penalties could lead to greater long-term rewards). The differences in games seem important enough to warrant research in dynamic parameter adjustment for better flexibility, possibly in the form of a meta-heuristic that analyses the game being played (or even more granular, at game state level), and choosing the right configuration for the given scenario.

A first step in this direction was taken recently by Gaina, Lucas, and Pérez Liébana [2019]. The approach taken in this study is at game state level, analyzing the features observed by the agent about its decision making process in a given state in order to dynamically adjust the length of the individuals. In this case, only one feature was used, how flat the landscape fitness looked to the agent after all its simulations, and the length adjusted as per Algorithm 3.4, with frequency $\omega = 15$, a lower bound $SD_- = 0.05$ (which indicates the length will be increased if fitness landscape flatness value falls below this value) and an upper bound $SD_+ = 0.4$ (which indicates the length will be decreased if fitness landscape flatness value rises above this value). The fitness landscape is represented by a collection of all fitness values observed during one game tick; the flatness value then becomes the standard deviation (δ) of all fitness values. This method therefore tries to adjust the length so that more individuals can be sampled if the landscape is varied, in order to gather enough statistics to make correct decisions; or so that long-term rewards can be found more easily with longer rollouts, if the fitness landscape is flat. It was shown to increase performance in sparse reward games for MCTS (while not affecting dense reward games), but it turned out to be detrimental for RHEA and halve its performance instead. This was thought to be due to the shift buffer enhancement (see next section), which is not compatible with dynamically adjusted sequence lengths.

3.5.2   RHEA ENHANCEMENTS

As we've seen in the beginning of this section on Rolling Horizon Evolution, vanilla RHEA does not sample the search space efficiently enough to be able to find good solutions in the short budget allocated for real-time decision computation. This subsection therefore explores ways to improve upon the base algorithm by incorporating other techniques or even by combining RHEA with other algorithms for interesting and high-performing hybrids.

Population initialization   A first step toward vanilla RHEA improvement was taken in a 2017 study by Gaina, Lucas, and Pérez Liébana [2017a], which looked at the very first step of the algorithm described in Section 3.5.1: population initialization. The theory behind it is that since RHEA cannot find a good enough solution quickly starting from random individuals, it would make sense that starting from an initially good solution could lead to better results.

There are many ways explored in literature to initialize evolutionary algorithms, although little for the specific application we have at hand here. Kazimipour, Li, and Qin [2014] present a nice review of such initialization methods for evolutionary algorithms, looking at the randomness of the method, its generality or compositionality. Even though some of the methods described are hinted at working well within general settings as proposed in GVGAI, they are also noted to be computationally expensive and not directly applicable to real-time games. As we are interested in performance boosts within limited time budgets, these might not be the best option.

Research in the area is encouraged, however, by earlier research in the game Othello [Kim, Choi, and Cho, 2007], which, even though still not real-time, showed significant improvement when the EA is initialized with an optimal solution determined by Temporal Difference Learning.

Algorithm 3.4 Adjusting rollout length dynamically.

Input: t: current game tick
Input: Ld: the fitness landscape (all fitness values) observed in the previous game tick
Input: L: rollout length
Requires: ω: adjustment frequency
Requires: SD_−: lower f_{Ld} limit for L increase
Requires: SD_+: upper f_{Ld} limit for L decrease
Requires: M_D: rollout length modifier
Requires: MIN_L: minimum value for L
Requires: MAX_L: maximum value for L

1: **procedure** DYNLENGTH(Ld, t, L)
2:   **if** $t \bmod \omega = 0$ **then**
3:     **if** $Ld = null$ **then**
4:       $f_{Ld} \leftarrow SD_-$ {f_{Ld} is a measure of the fitness landscape flatness}
5:     **else**
6:       $f_{Ld} \leftarrow \delta(Ld)$ {get standard deviation}
7:     **end if**
8:     **if** $f_{Ld} < SD_-$ **then**
9:       $L \leftarrow L + M_D$
10:     **else if** $f_{Ld} > SD_+$ **then**
11:       $L \leftarrow L - M_D$
12:     **end if**
13:     BOUND(L, MIN_L, MAX_L) {sequence length capped between min and max}
14:   **end if**
15:
16: **procedure** BOUND(L, MIN_L, MAX_L)
17:   **if** $L < MIN_L$ **then**
18:     $L \leftarrow MIN_L$
19:   **else if** $L > MAX_L$ **then**
20:     $L \leftarrow MAX_L$
21:   **end if**
22: **return** L

Following this line of research, Gaina, Lucas, and Pérez Liébana [2017a] use two different algorithms to produce initial optimal solutions from which to start the evolutionary process: a One Step Look Ahead greedy algorithm (1SLA, which chooses simply the action which leads to the next best state in any given state) and MCTS. These algorithms are given a chunk of RHEA's thinking budget (half for MCTS, enough to produce one individual for 1SLA) to return one good solution, which becomes the first individual in the population. This first solution is then mutated to form the rest of the initial population, and evolution proceeds as before.

The effects of these seeding options were tested using different values for the P and L parameters, as these affect not only the number of generations RHEA can perform, but also how much budget the seeding algorithms have to be allocated in order to generate a full individual, as well as how much this initial individual is disturbed (most in high population sizes). Results showed that, generally, MCTS seeding leads to a significantly better performance, although it is also significantly worse in four of the games in which MCTS typically performs poorly (although in this form, MCTS-seeded RHEA still performs better than simply MCTS). This could indicate that RHEA is unable to change the initial solution provided enough to fully account for the weaknesses of MCTS, thus higher mutation rates or different operators could be needed in order to make the best of both algorithms.

The 1SLA seeding appeared to be detrimental in most cases, possibly due to the fact that RHEA was unable to escape the initial local optima provided through seeding. There were games, however, where even this method was better than vanilla RHEA, suggesting that dynamically changing the seeding method depending on game type (or, as seen before, more granular at state level) could significantly improve results in specific games, as well as on average.

Bandit-based mutation   Since the mutation operator appeared to be one of the problems in the seeding-based methods described previously, it is natural to explore alternatives. One option that showed promise in other work [Liu, Pérez Liébana, and Lucas, 2017a] was a bandit-based mutation operator: this uses two levels of multi-armed bandit systems, one at individual level to choose which gene to mutate, and another at gene level to choose the new value given to the gene. Both of the systems employ the UCB equation (Equation (3.1)) with a constant $C = \sqrt{2}$ in order to balance between exploration of potentially good values and exploitation of known good mutations. The $Q(s, a)$ values are updated based on the new individual fitness, including the option to revert the mutation if this proved to be detrimental to the action sequence, aiming to always improve individuals.

Although this worked in previous applications of the method, it performed very poorly overall in the tests performed by Gaina, Lucas, and Pérez Liébana [2017b] on GVGAI games, in most cases being worse than vanilla RHEA, even when increasing the population size and individual length to get the best performance out of this hybrid. The bad performance is most likely due to the fact that changing one gene in the middle of the action sequence affects the meaning of the following actions as well—and it could be that one of the other actions were the ones producing the change in fitness wrongly attributed to the mutated gene instead—therefore a better calculation of the value of the mutation could potentially improve the performance of this enhancement. This focused mutation for improvement also has the potential of getting the algorithm stuck in local optima.

Statistical tree   Similar to the work in Pérez Liébana et al. [2015a] which showed promise, it's possible to keep more statistics throughout the evolutionary process. Adopting the way MCTS computes statistics about the actions it explores, but without relying the search on these statistics (so the search would still be performed by the regular evolutionary algorithm previously described), is an interesting way of deciding which action to finally play. This final decision would be based in this case on the action at the root of the tree with the highest UCB value, instead of the first in the best plan evolved. Figure 3.7 shows how the actions would be stored in the statistical tree after every individual rollout, using the individual fitness to backup the values throughout all the actions.

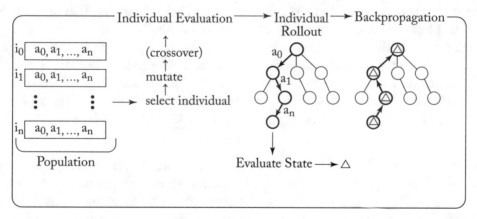

Figure 3.7: RHEA statistical tree steps.

This enhancement did work fairly well in the tests in Gaina, Lucas, and Pérez Liébana [2017b], performing the best (and better than vanilla RHEA) in low configurations for the parameters P and L, thus when enough individuals were evaluated to build significant statistics in the tree. In high configurations the algorithm was unable to gather enough statistics, suggesting it is better to play the first action of the best overall plan evolved instead. This could suggest further applications in dynamic parameter configuration work: switching this enhancement on when parameter values are low enough could lead to further boost in performance (e.g., compared to the work described in Gaina, Lucas, and Pérez Liébana [2019]).

Shift buffer   The next enhancement we're going to discuss is partly related to the first step of the algorithm as well (population initialization, that is). We say, partly, because it does not actually form a new initial population. This is instead a method of keeping the evolved population of individuals between game ticks, instead of starting from new random points each time.

A shift buffer refers to the technique through which all individuals in the final population in one game tick become the individuals in the first population in the next game tick. However, because the first action in the sequences evolved was played, we need to shift the search horizon to start from the second action in each individual and bring them up to date to the current game time. Therefore, the first action in all individuals is removed, while a new random action is added at the end, in order to keep the same length of individuals. In order to make sure all actions are still legal in case a change in the action space occurred during game ticks, any illegal actions are replaced with new random actions as well.

This enhancement showed promise when applied to MCTS within the context of the Physical Travelling Salesman Problem [Pérez Liébana et al., 2015a] and, even though it does not appear to work as well for MCTS in GVGAI, it does help RHEA make better use of limited thinking budgets by reusing already evolved populations and getting to improve the plans further instead of discarding all of its computed information.

In the study by Gaina, Lucas, and Pérez Liébana [2017b], the addition of this enhancement led to a high win rate increase (as well as significantly higher scores) over vanilla RHEA in all configurations. The shift buffer is not the best combination with bandit mutation, possibly due on one hand to the general poor performance of bandit variations, but also due to the old statistics used by the bandits to make their recommendations. The success of this variant encouraged its use in several other works, such as that by Santos, Bernardino, and Hauck [2018] who use a Rolling Horizon Evolutionary Algorithm with a shift buffer for General Game Playing, with good results.

Monte Carlo rollout evaluation   One last enhancement that's been explored in research [Horn et al., 2016] involves step 2 of the vanilla RHEA algorithm (individual evaluation) and is inspired by Monte Carlo rollouts as in MCTS. The idea here is that after finishing the usual advancing through the actions in the sequence, we add further horizon to the search with a random rollout of length r_L, possibly sampled multiple times, r_N, so that the results are more significant. In this case, the fitness of the individual is instead given by the average values of states reached after r_L more actions executed at the end of the action sequence, as in Equation (3.10):

$$f = \frac{\sum_{n=1}^{r_N} V(s_n)}{r_N}. \tag{3.10}$$

As opposed to the case where the sequence length L is increased directly, this variant offers the possibility of exploring more varied (and not fixed) action sequences, which allows it to find interesting variations in the search space it might not otherwise. In the study by Gaina, Lucas, and Pérez Liébana [2017b], r_L is given a value as half the length of the individual ($L/2$), while r_N is tested for different values: 1, 5, and 10. This variant was tested individually against vanilla RHEA, but also in combination with previously described enhancements (with the exception of the bandit-based mutation, which was considered to be performing too poorly to consider for this last experiment). An overview of these results can be observed in Figure 3.8; MCTS is also included in the figure for comparison. A further summary of the best variants obtained in all configuration of parameters P and L can also be observed in Table 3.5.

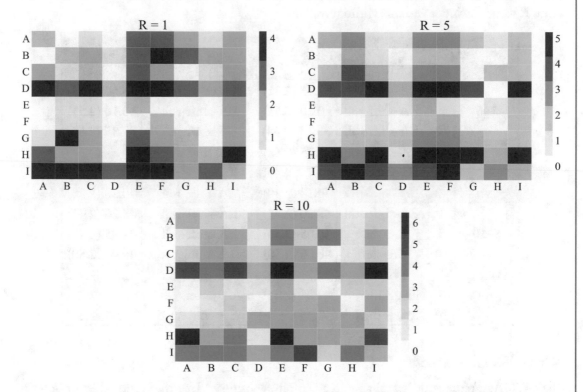

Figure 3.8: Win percentage for configuration 10–14. The color bar denotes in how many unique games row was significantly better than column. Legend: A = Vanilla, B = EA-roll, C = EA-shift, D = EA-shift-roll, E = EA-tree, F = EA-tree-roll, G = EA-tree-shift, H = EA-tree-shift-roll, I = MCTS.

Overall, Monte Carlo rollouts at the end of individual evaluation appeared to offer a nice boost in performance, especially when combined with a shift buffer, significantly outperforming vanilla RHEA (likewise, the shift buffer enhancement is even better if in combination with MC rollouts). As may be expected, however, MC rollouts work best if P and L are lower, due to the limited budget in higher configurations that would now have to be split between action sequences and MC rollouts as well. These computations become fairly expensive as individual length grows, while still yielding good results.

Regarding r_N values, it appeared that 5 was best in many cases. A highlight of this is given by the variant combining the shift buffer and MC rollouts, which matches the performance of MCTS when $r_N = 5$.

Table 3.5: The best algorithms (by Formula-1 points and win rate) in all configurations and rollout repetitions (R), as compared against the other variants in the same configuration and the same R value (includes variants without rollouts)

Config.	R	Best By F1 Points		Best By Win Rate	
		Algorithm	Avg. Wins	Algorithm	Avg. Wins
1–6	1	EA-shift-roll	38.35 (2.31)	EA-tree-shift-roll	38.60 (2.55)
	5	EA-shift-roll	40.10 (2.51)	EA-shift-roll	40.10 (2.51)
	10	EA-shift-roll	39.35 (2.64)	EA-shift-roll	39.35 (2.64)
2–8	1	EA-shift-roll	40.35 (2.63)	EA-shift-roll	40.35 (2.63)
	5	EA-shift-roll	40.75 (2.46)	EA-shift-roll	40.75 (2.46)
	10	EA-shift-roll	40.20 (2.30)	EA-shift-roll	40.20 (2.30)
5–10	1	EA-shift-roll	43.20 (2.43)	EA-shift-roll	43.20 (2.43)
	5	EA-shift	40.05 (2.50)	EA-shift-roll	41.85 (2.42)
	10	EA-shift	40.05 (2.50)	EA-shift	40.05 (2.50)
10–14	1	EA-shift	39.75 (2.54)	EA-shift-roll	42.80 (2.44)
	5	EA-shift-roll	42.05 (2.48)	EA-tree-shift-roll	42.70 (2.41)
	10	EA-shift-roll	42.35 (2.53)	EA-shift-roll	42.35 (2.53)

3.6 EXERCISES

The GVGAI Framework is available in a Github repository.[3] Use the release 2.3[4] in order to run the same version presented in this chapter.[5]

3.6.1 MONTE CARLO TREE SEARCH

Tweaking MCTS

Execute the sample MCTS agent in the single-player planning setting and tweak its parameters to observe changes in performance.

- The code of this agent is in the package `tracks.singlePlayer.advanced.sampleMCTS`. In this package, `TreeNode.java` has several parameters that can be tweaked, such as the rollout length or the value of C in the UCB1 equation.

- The function to evaluate states at the end of a Monte Carlo simulation is `value()`. Try to create different value functions that can achieve better results than the default one (spoiler alert: it's not easy!).

[3]https://github.com/GAIGResearch/GVGAI
[4]https://github.com/GAIGResearch/GVGAI/releases/tag/2.3
[5]These exercises are also available at this book's website: https://gaigresearch.github.io/gvgaibook/.

Working with Knowledge-Based Monte Carlo Tree Search
The following ideas could be taken as exercises or research projects.

- Rather than using Euclidean distances for calculating the distances to sprites, implement and use a path-finding technique instead. Distances will be more accurate at the expense of computational cost. How could you maximize the impact of the former and minimize that of the latter?

- Substitute the simple $1 + 1$-EA algorithm used to evolve the weights during the Monte Carlo simulation for a more involved evolutionary technique.

- Can you think of better features (than distances to sprites) that become better guides for KB-MCTS? Maybe some convolutional filters over the space around the avatar or the whole game state?

- Expand the experiments done with KB-MCTS to the other game sets of the GVGAI framework.

Expanding the Multi-Objective Monte Carlo Tree Search
The following ideas could be taken as exercises or research projects.

- All approaches run in the experiments described in Section 3.4 use a determined set of fixed weights. You can experiment to see what happens if these weights are different. Could you find a way to dynamically change the weights in response to the state of the game and improve the performance?

- The mixed strategy approach showed some promising results and it is possible that a high performance can be obtained if the policy for switching objectives is optimized. How would you do this? A possibility is to add game and agent related features (see Section 4.4) to this policy for objective decision-making.

- Explore other multi-objective optimization approaches. For instance, the epsilon-constrained approach [Miettinen, 1999], which considers one of the objective functions as a constraint ($O_i < \epsilon$, in case of maximization) and only the other function is subject to optimization.

3.6.2   ROLLING HORIZON EVOLUTIONARY ALGORITHMS
Tweaking RHEA
Execute the sample RHEA agent in the single-player planning setting and tweak its parameters to observe changes in performance.

- The code of this agent is in the package `tracks.singlePlayer.advanced.sampleRHEA`. In this package, `Agent.java` has several parameters that can be tweaked, such as the population size, individual lengths and evolutionary operators.

- The function to evaluate an individual is `evaluate()`. Try to create different value functions that can achieve better results than the default one.

Working with Rolling Horizon Evolutionary Algorithms

The following ideas could be taken as exercises or research projects.

- Tweak the parameters of the Dynamic Rollout Length adjustment. Can this be change dynamically in response to features of the game state?

- What other algorithms could provide good initial solutions to seed the initial population of RHEA? What is the effect of changing the budget for the algorithm that seeds the population (i.e., what happens if the allotted budget is less than 50% of the total time for MCTS)?

- Also on RHEA seeding, you can investigate how much should the solution provided by seeding algorithms be disturbed in order to form the initial population.

- Try improving the bandit based mutation. This mutation is univariate (only one gene is considered at a time), but it is likely that higher-dimensional tuples provide better results if they are able to identify the relationships between the genes in the sequence.

CHAPTER 4

Frontiers of GVGAI Planning

Diego Pérez Liébana and Raluca D. Gaina

4.1 INTRODUCTION

Multiple studies have tackled the problem presented in the planning track of GVGAI. This chapter aims to present the state-of-the-art on GVGAI, describing the most successful approaches on this domain in Section 4.2. It is worth highlighting that these approaches are based on different variants of tree search methods, predominantly MCTS (see previous chapter), with the addition of several enhancements or heuristics. These agents submitted to the competition have shown the ability of playing multiple games minimizing the amount of game-dependent heuristics given, truly reflecting the spirit of GVGAI.

It is important to observe and analyze these methods because, despite being the most proficient ones, they are still far from an optimal performance on this problem. On average, these approaches achieve around 50% of victory rate on games played. From the study of the limitations of these methods one can build stronger algorithms that tackle this challenge better. Section 4.3 summarizes the open challenges that must be faced in order to make progress in this area.

One of the most promising options is to automatically extract information from the game played in an attempt to *understand* it. Different games may benefit from distinct approaches, and being able to identify when to use a determined algorithm or heuristic can provide a big leap forward. In a similar vein, being able to understand the situation of the agent in the game (i.e., is the agent on the right track to win?) can provide automatic guidelines as to when to change the playing strategy. Section 4.4 of this chapter describes our approach to predict game outcomes from live gameplay data [Gaina, Lucas, and Pérez Liébana, 2018]. This is meant to be another step toward a portfolio approach that improves the state of the art results in the planning challenge of general video game playing.

GVGAI planning is a hard problem, and we have reached a point where general agents can play decently some of the available games. The next step is maybe the hardest (but arguably the most interesting one): can we automatically analyze games and determine what needs to be done to win? How can we understand the goals of any game and how closely is the agent to achieve them? The answers to these questions may lead to a new generation of general video game playing agents.

4.2 STATE OF THE ART IN GVGAI PLANNING

This section presents three agents that have won several editions of the GVGAI competitions: OLETS, ToVo2 and YOLOBOT. We describe the methods implemented and the strengths and weaknesses of each approach.

4.2.1 OLETS (ADRIENCTX)

Open loop expectimax tree search (OLETS) is an algorithm developed by Adrien Couëtoux for GVGAI that won the first edition of the single-player GVGAI competition. Furthermore, its adaptation to the 2-player planning track [Gaina et al., 2017a] was also successful at winning two editions of such track (see Table 2.1).

OLETS takes its inspiration from Hierarchical Open-Loop Optimistic Planning (HOLOP) [Weinstein and Littman, 2012], which is uses the FM to sample actions in a similar way to MCTS, without ever storing the states of the tree in memory. Although this is not an optimal approach for non-deterministic domains, it works well in practice. One key difference with HOLOP is that OLETS does not use the default UCB policy for action selection (Equation (3.1)) but a new method called *Open Loop Expectimax* (OLE). OLE uses $r_M(n)$ instead of the average of rewards in the policy, which is the linear combination of two components: the empirical average reward computed from the simulations that went through the node n and the maximum r_M value among its children. Another key difference is that OLETS does not perform any rollouts. Equation (4.1) shows the OLE score, where $n_s(n)$ is the number of times the state s has been visited and $n_s(a)$ the number of times action a has been chosen from state s:

$$score = r_M(a) + \sqrt{\frac{\ln(n_s(n))}{n_s(a)}}. \tag{4.1}$$

Algorithm 4.5 describes OLETS in detail. OLETS tries to reward exploration of the level via a *taboo bias* added to the state value function. When new nodes are added to the tree, their state receives a penalty (of an order of magnitude smaller than 1) if it has been visited previously less than T steps ago. Values of $10 < T < 30$ has shown to provide good empirical results.

One of the weaknesses of OLETS is the absence of learning. It's a very simple and flexible method that gives the agent good reactive capabilities, mainly due to the open-loop search (in contrast to closed-loop that stores the states of the games in the node). However, it struggles in games where long computations are needed to solve a game that requires a deeper planning.

4.2.2 TOVO2

The ToVo2 controller is an agent developed by Tom Vodopivec and inspired in MCTS and RL [Sutton and Barto, 1998]. ToVo2 was the first winner of the 2-player GVGAI planning competition.

Algorithm 4.5 OLETS, from Pérez Liébana et al. [2015]. $n_s(n)$: number of simulations that passed through node n; $n_e(n)$: number of simulations ended in e; $R_e(n)$: cumulative reward from simulations ended in n; $C(n)$: set of children of n; $P(n)$: parent of n.

Input: s: current state of the game.
Input: T: time budget.
Output: action: to play by the agent in the game.

1: **procedure** OLETS (s, T)
2:     $\mathcal{T} \leftarrow root$ {initialize the tree}
3:     **while** elapsed time $< T$ **do**
4:         RunSimulation$(\mathcal{T}, s)$
5:     **end while**
6:     **return** action $= \operatorname{argmax}_{a \in C(root)} n_s(a)$
7:
8: **procedure** RunSimulation $(\mathcal{T}, s_0)$
9:     $s \leftarrow s_0$ {set initial state}
10:     $n \leftarrow root(\mathcal{T})$ {start by pointing at the root}
11:     $Exit \leftarrow$ False
12:     **while** $\neg Final(s) \wedge \neg Exit$ **do**
13:         **if** n has unexplored actions **then**
14:             $a \leftarrow$ Random unexplored action
15:             $s \leftarrow$ ForwardModel(s,a)
16:             $n \leftarrow$ NewNode(a,Score(s))
17:             $Exit \leftarrow$ True
18:         **else**
19:             $a \leftarrow \operatorname{argmax}_{a \in C(n)} \text{OLE}(n,a)$ {select a branch with OLE (Equation (4.1))}
20:             $n \leftarrow a$
21:             $s \leftarrow$ ForwardModel(s,a)
22:         **end if**
23:     **end while**
24:     $n_e(n) \leftarrow n_e(n) + 1$
25:     $R_e(n) \leftarrow R_e(n) + Score(s)$
26:     **while** $\neg P(n) = \emptyset$ **do**
27:         $n_s(n) \leftarrow n_s(n) + 1$ {update the tree}
28:         $r_M(n) \leftarrow \frac{R_e(n)}{n_s(n)} + \frac{(1 - n_e(n))}{n_s(n)} \max_{c \in C(n)} r_M(c)$
29:         $n \leftarrow P(n)$
30:     **end while**

ToVo2 enhances MCTS by combining it with Sarsa-UCT(λ) [Vodopivec, Samothrakis, and Ster, 2017] to combine the generalization of UCT with the learning capabilities of Sarsa. The algorithm is also an open-loop approach that computes state-action values using the Forward Model during the selection and simulation phases of MCTS. Rewards are normalized to [0, 1] to combine the UCB1 policy (Equation (3.1)) with Sarsa. Other parameters are an exploration rate $C = \sqrt{2}$, a reward discount rate $\gamma = 0.99$ and the eligibility trace decay rate $\lambda = 0.6$.

The simulation step computes the value of all states visited, instead of only observing the state found at the end of the rollout. This value is calculated as the difference in score between two consecutive game ticks. All visited nodes in an iteration are retained and 50% of knowledge is forgotten after each search due to the updated step-size parameter. The opponent is modeled as if it were to move uniformly at random (i.e., it is assumed to be part of the environment).

Two enhancements for the MCTS simulation phase augment the controller. The first one is weighted-random rollouts, which bias the action selection to promote exporation and visit the same state again less often. The second one is a dynamic rollout length, aimed at an initial exploration of the close vicinity of the avatar (starting with a depth of 5 moves from the root) to then increase the depth progressively (by 5 every 5 iterations, up to a maximum of 50) to search for more distant goals.

Similar to OLETS, ToVo2 struggles with games that see no rewards in the proximity of the player, but it is robust when dealing with events happening in its close horizon. In games where the exploration of the level is desirable, the weighted-random rollouts proved to be beneficial, but if offered no advantage in puzzle games where performing an exact and particular sequence of moves is required.

4.2.3 YOLOBOT

A submitted agent that is worth discussing is YOLOBOT, which won three editions of the GVGAI single-player planning track (outperforming OLETS in them). YOLOBOT was developed by Tobias Joppen, Miriam Moneke and Nils Schröder, and its full description can be found at Joppen et al. [2018].

YOLOBOT is a combination of two different methods: a heuristic-guided Best First Search and MCTS. The former is used in deterministic games, while the latter is employed in stochastic ones. The heuristic search attempts to find a sequence of moves that can be replicated exactly. At the start, the agent returns the NIL action until this sequence is found, a limit of game ticks is reached or the game is found to be stochastic. In the latter case, the agent switches to an enhanced MCTS to play the game. This algorithm is enhanced in several ways.

- Informed priors: YOLOBOT keeps and dynamically updates a knowledge base with predictions about events and movements within the game. This knowledge base is used in an heuristic to initialize the UCT values for non-visited nodes, biasing MCTS to expand first through those actions that seem to lead to more promising positions. These values are disregarded as soon as MCTS starts a simulation from them.

- Informed rollout policies: the same heuristic used to find promising positions is employed to bias action selection during the simulation phase of MCTS.

- Backtracking: when finding a losing terminal state, the algorithm backtracks one move and simulates up to four alternative actions. If one of these alternative moves leads to a non-losing state, the value of this one is backpropagated instead.

- Pruning of the search space: the FM is used to determine if the next state of s when applying action a is the same as the state reached when NIL is played. In this case, all actions a that meet this criteria are pruned. Other actions leading the avatar outside the level bounds and those that lead to a losing game state are also pruned.

YOLOBOT has achieved the highest win rate in GVGAI competitions since its inception. This agent works well in deterministic games (puzzles) where the solution is reachable in a short to medium-long sequence of actions, and it outperforms the other approaches when playing stochastic games. This, however, does not mean that YOLOBOT excels at all games. As shown in Table 2.1, YOLOBOT has achieved between 41.6% and 63.8% of victories in different game sets of the framework. There is still about 50% of the games where victory escapes even to one of the strongest controllers created for this challenge.

4.3   CURRENT PROBLEMS IN GVGAI PLANNING

Both planning tracks of GVGAI (single and 2-player) have received most attention since the framework was built and made public. As hinted in the previous section and shown in Table 2.1 in Chapter 2, the best approaches do not achieve a higher than 50% winning rate. Furthermore, many games are solved in very rare cases and the different MCTS and RHEA variants struggle to get more than 25% of victories in all (more than 100) games of the framework. Thus, one of the main challenges at the moment is to increase the win percentage across all games of the framework.

A recent survey [Pérez Liébana et al., 2018b] of methods for GVGAI describes many enhancements for algorithms that try to tackle this issue. In most cases, including the ones described in this book, the improvements do achieve to increase performance in a subset of the games tried, but there is normally another subset in which performance decreases or (in the best case) stays at the same level. The nature of GVGP makes this understandable, as it is hard to even think of approaches to work well for all games—but that does not mean this is not a problem to solve.

This recent survey shows, however, a few lines of work that could provide significant advances in the future. For instance, it seems clear that feature extraction from sprites work better if using a more sophisticated measure (not only a straightforward A*, but also other methods like potential fields [Chu, Harada, and Thawonmas, 2015]) than simple Euclidean distances. How to combine this more complex calculations with the real-time aspect of the games, especially in those games where a wise use of the budget is crucial, is still a matter for future investigation.

The wise use of the budget time is an important point for improvement. There is a proliferation of methods that try to use the states visited with the FM during the thinking time to learn about the game and extract features to bias further searches (as in Pérez Liébana, Samothrakis, and Lucas [2014] and Joppen et al. [2018]). In most cases, these approaches work well providing a marginal improvement in the overall case, but the features are still tailored to a group of games and lack generality. Put simply, when researchers design feature extractors, they are (naturally) influenced by the games they know and what is important in them, but some other games may require more complex or never-thought-before features. The design of an automatic and general feature extractor is one of the biggest challenges in GVGP.

Another interesting approach is to work on the action space to use more abstract moves. This can take the for of macro-actions (sequences of atomic actions treated as a whole [Pérez Liébana et al., 2017]) or options (in MCTS [Waard, Roijers, and Bakkes, 2016], associated with goals). This approach makes the action space coarser (reduces the action space across several consecutive turns) and maximizes the use of the budget times: once a macro-action starts its execution (which takes T time steps to finish), the controller can plan for the next macro-action, counting on $T - 1$ time steps to complete that search. This can be an interesting approach for games that require long-term planning, a subset of games that has shown to be hard for the current approaches. Results of using these action abstractions show again that they help in some games, but not in others. They also suggest that some games benefit from different sets of lengths of macro-actions, while other games are better played with others.

It is reasonable to think that, given that certain algorithms perform better at some games than others, and some games are played better by different methods, an approach that tries to automatically determine what is the right algorithm for the right game should be of great help. In fact, the already discuss YOLOBOT agent takes a first stab at this, by distinguishing between deterministic and stochastic games. Game classification [Bontrager et al., 2016, Nelson, 2016] and the use of hybrids or hyper-heuristic methods are in fact an interesting area for research. The objective would be to build a classifier that dynamically determines which is the best algorithm to use in the current game and then switch to it. Some attempts have been made to classify games using extracted features, although the latest results seem to indicate that these classifications (and the algorithms used) are not strong enough to perform well across many games.

One interesting direction that we start exploring in the next chapter is the use of more general features, focused on how the agent *experiences* the game rather than features that are intrinsic to it (and therefore biased). The next study puts the first brick in a system designed to switch between algorithms in game based only on agent game-play features. In particular, the next section describes how a win predictor can be built to determine the most likely outcome of the game based on agent experience.

4.4   GENERAL WIN PREDICTION IN GVGAI

The objective of this work is to build an outcome predictor based only on measurements from the agent's experience while playing any GVGAI game. We purposefully left game-related features (like the presence of NPCs or resources) outside this study, so the insights can be transferred to other frameworks or particular games without much change. The most important requirements are that the game counts on an FM, a game score and game end states with a win/loss outcome.

4.4.1   GAME PLAYING AGENTS AND FEATURES

The first step for building these predictors is to gather the data required to train the predicting models. This data will be retrieved from agents playing GVGAI games. The algorithms used will be as follows.

Random Search (RS)

The RS agent samples action sequences at random of length L until the allocated budget runs out. Then, the first action of the best sequence is played in the game. Sequences are evaluated using the FM to apply all actions from the current game state and assigning a value to the final state following Equation (4.2). H^+ is a large positive integer (and H^- is a large negative number):

$$f = score + \begin{cases} H^+, & \text{if loss} \\ H^-, & \text{if win.} \end{cases} \tag{4.2}$$

Three configurations for RS are used in this study, according to the value of L: 10, 30, and 90.

Rolling Horizon Evolutionary Algorithm (RHEA)

The RHEA agent (explained in Section 3.5) is used for this study, using some of the improvements that have shown to work well in previous studies. The selected algorithm configurations are:

- Vanilla RHEA [Gaina et al., 2017b], the default configuration of the algorithm.

- EA-MCTS [Gaina, Lucas, and Pérez Liébana, 2017a], in which the original population is seeded by MCTS.

- EA-Shift [Gaina, Lucas, and Pérez Liébana, 2017b], in which RHEA is enhanced using the Shift Buffer and Monte Carlo rollouts at the end of the individuals.

- EA-All, which combines EA-Shift with EA-MCTS, for completeness.

For these four variants, two different parameters sets where used: $P = 2$, $L = 8$ and $P = 10$, $L = 14$, where L is individual length and P population size. Final states are evaluated using Equation (4.2).

Monte Carlo Tree Search (MCTS)

The vanilla MCTS algorithm is used, using again Equation (4.2) to evaluate states reached at the end of the 3 parameter sets were used for MCTS: $W = 2$, $L = 8$; $W = 10$, $L = 10$, and $W = 10$, $L = 14$. W is the number of MCTS iterations and L the depth of the rollouts from the root

All these 14 algorithms were run in the 100 public games of the GVGAI framework, using the 5 levels available per game and 20 repetitions per level. All algorithms counted on the same budget per game tick to make a decision: 900 calls to the FM's advance function. Table 4.1 summarizes the results obtained by these methods in the 100 games tested.

Table 4.1: Victory rate (and standard error) of all methods used in this study across 100 GVGAI games. Type and configuration (rollout length L if one value, population size P and roll-out length L if two values) are reported.

#	Algorithm	Victory Rate (Standard Error)
1	10-14-EA-Shift	26.02% (2.11)
2	2-8-EA-Shift	24.54% (2.00)
3	10-RS	24.33% (2.13)
4	14-MCTS	24.29% (1.74)
5	10-MCTS	24.01% (1.65)
6	10-14-EA-MCTS	23.99% (1.80)
7	2-8-EA-MCTS	23.98% (1.73)
8	2-8-EA-ALL	23.95% (1.98)
9	8-MCTS	23.42% (161)
10	10-14-RHEA	23.23% (2.08)
11	10-14-EA-ALL	22.66% (2.02)
12	30-RS	22.49% (2.02)
13	2-8-RHEA	18.33% (1.77)
14	90-RS	16.31% (1.67)

Each one of these runs produced two log files with agent and game state information at every game state. Regarding game, the score at each step and the final game result (win/loss) are saved. Regarding the agent, we logged the action played at each game step and the set of features described as follows.

• ϕ_1 **Current game score**.

- ϕ_2 **Convergence**: the iteration number when the algorithm found the final solution recommended and did not change again until the end of the evolution, during one game step. A low value indicates quick and almost random decisions.

- ϕ_3 **Positive rewards**: the count of positive scoring events.

- ϕ_4 **Negative rewards**: the count of negative scoring events.

- ϕ_5 **Success**: the slope of a line over all the win counts. This count reflects the number of states which ended in a win at any point during search. A high value shows an increase in discovery of winning states as the game progresses.

- ϕ_6 **Danger**: the slope of a line over all the loss counts. This count grows for every end game loss found during search. A high value indicates that the number of losing states increases as the game progresses.

- ϕ_7 **Improvement**: given the best fitness values seen since the beginning of the game, improvement is the slope of a this increment over game tick. A high value indicates that best fitness values increase as the game progresses.

- ϕ_8 **Decisiveness**: this is the Shannon Entropy (SE, see Equation (4.3)) over how many times each action was recommended. In all cases where the feature is calculated as SE, a high value suggests actions of similar value; the opposite shows some of these actions to be recommended more often.

- ϕ_9 **Options exploration**: SE over the number of times each of the possible actions was explored. In this case, this reflects how many times this action was the first move of a solution at any time during search. A low value shows an imbalance in actions explored while a high value means that all actions are explored approximately the same as first moves during search.

- ϕ_{10} **Fitness distribution**: SE over fitness per action.

- ϕ_{11} **Success distribution**: SE over win count per action.

- ϕ_{12} **Danger distribution**: SE over loss count per action.

$$H(X) = -\sum_{i=0}^{N-1} p_i \log_2 p_i. \tag{4.3}$$

Features ϕ_2, ϕ_8, ϕ_9, ϕ_{10}, ϕ_{11}, and ϕ_{12} compute averages from the beginning of the game up until the current tick t. Features ϕ_5, ϕ_6, ϕ_{11}, and ϕ_{12} rely on the FM. Data set and processing scripts have been made publicly available.[1]

---

[1]https://github.com/rdgain/ExperimentData/tree/GeneralWinPred-CIG-18. The final data is split over 281,000 files. It took approximately 2.5 hours to generate this database from raw data files, using a Dell Windows 10 PC, 3.4 GHz, Intel Core i7, 16GB RAM, 4 cores.

Additionally, features are divided into 3 different game *phases*: early (first 30% ticks of the games), late (ticks from the last 70% of the game) and mid-game (ticks between 30–70% of the game). These divisions were used to train different phase models: early, mid and late game classifiers.

Feature correlation

The division in different game stages corresponds to the belief that different events generally occur at the beginning and end of the game. Therefore, having models trained in different game phases may produce interesting results. Figure 4.1 shows the correlation between features using the Pearson correlation coefficient. This compares the early (left) and late (right) game features showing small (but existing) differences. For instance, there are higher correlations between the features located in the bottom right corner in the late game than in the early game.

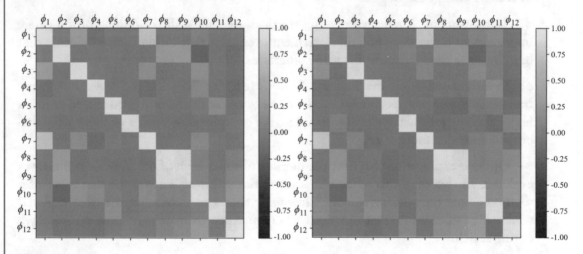

Figure 4.1: Feature correlation early game (left, 0–30% of all games) and late game (right, 70–100% of all games).

The late game phase also shows an interesting and positive correlation between convergence (ϕ_2) and danger (ϕ_6). This may suggest that agents take more time to settle on their final decisions when more possible losses are found within the search. A positive correlation that is less strong in the late game is that between fitness improvement and fitness distribution over the actions, implying that when one action is deemed significantly better than the rest, it is unlikely for the fitness to improve further, possibly due to the other actions not being explored further. Finally, a persistent negative correlation exists between convergence and fitness distribution. This seems to indicate that an agent, when finding an action that is deemed prominently better than the rest, it is not likely to change its decision.

4.4.2   PREDICTIVE MODELS

For all models trained for this study the data was randomly split in training (80 of the data) and test sets (20). Several classifier models are built using the features described as input and a win/loss as a label and prediction. Model prediction quality is reported in this section according to precision, recall and F1 score (Equations (4.4), (4.5), and (4.6), respectively):

$$precision = \frac{TP}{TP + FP},$$ (4.4)

$$recall = \frac{TP}{TP + FN},$$ (4.5)

$$F1 = 2 \cdot \frac{precision \cdot recall}{precision + recall},$$ (4.6)

where *TP* stands for true positives (correctly predicted a win), *FP* stands for false positives (incorrectly predicted a win) and *FN* stands for false negatives (incorrectly predicted a loss). F1 score is the main indicator reported due to the low overall performance of the general agents (close to 25% win rate in all games) and the high variety of the GVGAI games.

Baseline Model

In order to determine if the models trained are better than a simple expert rule system, a baseline model was built. In most games (GVGAI and other arcades), gaining score through the game is typically a good indicator of progression, and as such it normally leads to a victory. Equation (4.7) is a simple model that compares the count of positive and negative score events.

$$\hat{y} = \begin{cases} win & if \ \phi_3 > \phi_4 \\ lose & otherwise. \end{cases}$$ (4.7)

The performance of this classifier on the test set can be seen in Table 4.2. It can be observed that it achieves an an F1-Score of 0.59, despite having a high precision (0.70). This model is referred to as R_g in the rest of this chapter.

Table 4.2: Global rule based classifier report. Global model tested on all game ticks of all instances in the test set.

	Precision	Recall	F1-Score	Support
Loss	0.83	0.52	0.64	20,500
Win	0.35	0.70	0.46	7,500
Average/Total	**0.70**	**0.57**	**0.59**	**28,000**

Classifier Selection—Global Model

In this work, we trained and tested seven classifiers as a proof of concept for the win prediction task. These classifiers are K-Nearest Neighbors (5 neighbors), Decision Tree (5 max depth), Random Forest (5 max depth, 10 estimators), Multi-layer Perceptron (learning rate 1), AdaBoost-SAMME [Zhu et al., 2006], Naive Bayes and Dummy (a simple rule system, also used as another baseline). All classifiers use the implementation in the Scikit-Learn Python 2.7.14 library [Buitinck et al., 2013]. The parameters not specified above are set to their respective default values in the Scikit-Learn library.

Performance is assessed using cross-validation (10 folds). The classifiers obtained, following the same order as above, an accuracy of 0.95, 1.00, 0.98, 0.96, 1.00, 0.95, and 0.66 during evaluation. AdaBoost and Decision Tree achieved very high accuracy values in validation and test (see Table 4.3). The rest of the experiments use AdaBoost as the main classifier (with no particular reason over Decision Trees).

Table 4.3: AdaBoost tested on all game ticks of all instances in the test set

	Precision	Recall	F1-Score	Support
Loss	1.00	0.99	0.99	20,500
Win	0.97	0.99	0.98	7,500
Average/Total	**0.99**	**0.99**	**0.99**	**28,000**

Table 4.4 shows the importance of teach feature according to AdaBoost. Game score seems to be, unsurprisingly, the most important feature to distinguish wins and losses. This is closely followed by the number of wins seen by the agent, the improvement of the fitness and the measurement of danger. Decisiveness of the agent seems to have no impact in deciding the outcome of the game, according to the model trained.

Table 4.4: Importance of features as extracted from the global model

ϕ_1	0.24	ϕ_2	0.04	ϕ_3	0.08	ϕ_4	0.06
ϕ_5	0.2	ϕ_6	0.1	ϕ_7	0.12	ϕ_8	0
ϕ_9	0.06	ϕ_{10}	0.02	ϕ_{11}	0.02	ϕ_{12}	0.06

Model Training

Predictions were made at three different levels during the game: early game (first 30% of the game ticks), mid game (30–70%) and late game (last 30%). The models are trained using the features captured in the game ticks corresponding to each interval in the game. These three models are referred to as E_g, M_g, and L_g, respectively, in the rest of this section.

Trained models were tested by checking their performance on the 20 test games (also considering the game tick intervals). Training with 10-fold cross-validation provided 0.80, 0.82, and 0.99 as results for the early, mid and late game predictions, respectively. Test accuracies are 0.73, 0.80, and 0.99, with F1-Scores of 0.70, 0.80, and 0.99, respectively.

Live Play Results

Our next step was to simulate how the trained models predict the game outcome *live* (when an agent is playing). For this, we simulated play by extracting the features logged by agents in log files for a range of game ticks ($T = \{100 \cdot a : \forall a \in [1, 20] : a \in \mathbb{N}\}$), all from the beginning of the game until the current tick tested $t \in T$. We took games played by the 14 algorithms presented above on 20 test games, playing 20 times on each one of their 5 levels. Each model was asked to predict the game outcome every 100 ticks.

Figures 4.2 and 4.3 show the results obtained from this testing. The baseline rule-based model achieves a high performance in some games, showing to be better than the trained predictive models (i.e., see *Aliens*, *Defem*, *Chopper* and *Eggomania*). These games have plenty of scoring events, thus it is not surprising that the simple logic of R_g works well in these cases. However, there are other games in which the trained models perform much better predictions than the baseline (see *Ghost Buster*, *Color Escape* or *Frogs*), where the outcome is not significantly correlated with game score and the simple rule-based prediction is not accurate.

It is worthwhile mentioning that the trained models do not follow the expected curves. One could expect E_g performing better in the early game, to then decrease its accuracy when the game progresses. M_g could show high performance in the middle game and L_g offering good predictions only on the end game. However, the early game predictor has a worst performance compared to the rest (this could be explained by the lack of information available for this model). The late game model is very accurate in games with very low win rate (in *Fireman*, for instance, where E_g and M_g are predicting wins, yet the overall win rate remains at 0% for this game).

A high F1-score index indicate that the predictors are able to judge correctly both wins and losses. It is therefore interesting to pay attention to those games, like *Defem* and *Ghost Buster*, with close to 50% win rate. In these cases, F1-scores over 0.8 are achieved when only half the game has passed. The middle model M_g provides very good results in this situations, becoming the best predictor to use in this case and possibly the best one to use.

It is remarkable to see that the model is able to predict, half-way through the game (and sometimes just after only a fourth of the game has been played), the outcome of the game, even if games are won or lost with the probability as a coin flip. These models are general: they have been trained in different games without relaying in game-dependent features—just agent experience measurements.

Therefore, there is a great scope of using these predictors as part of a hyper-heuristic system. Some of the algorithm tested in this study *do* win at these 50% win rate games like *Defem* or *Ghost Buster* and finding a way to use the appropriate method for each game would boost

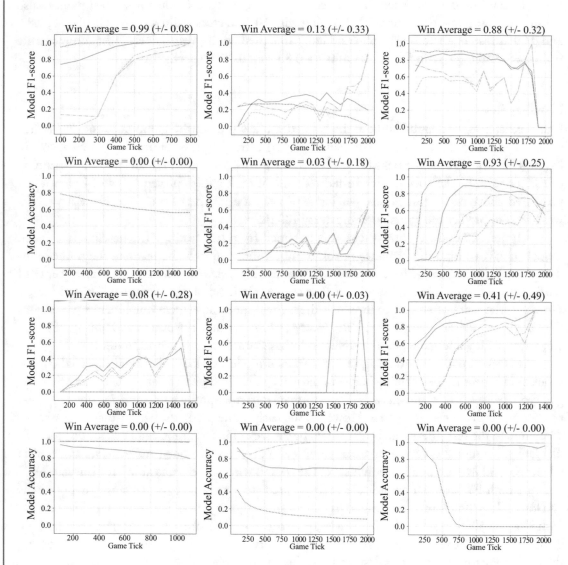

Figure 4.2: Model F1-scores for each game in the test set, averaged over up to 1,400 runs, 14 agents, 100 runs per game. Game ticks are displayed on the X axis, maximum 2,000 game ticks. three different predictor models trained on early, mid and late game data features, as well as the baseline rule-based predictor. If F1-scores were 0 for all models, accuracy is plotted instead. Additionally, win average is reported for each game. Games from top to down, left to right: *Aliens, Boulderdash, Butterflies, Caky Baky, Chase, Chopper, Color Escape, Cops, Defem, Deflection, DigDug* and *Donkeykong*.

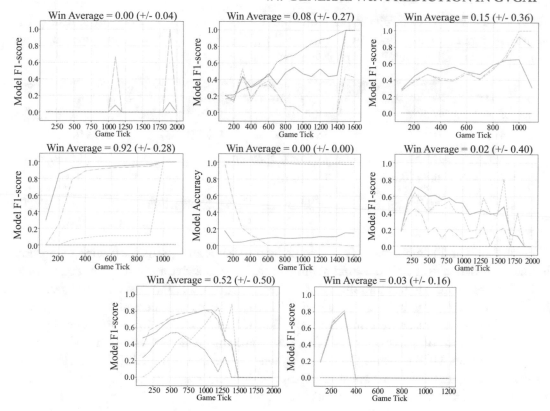

Figure 4.3: Model F1-scores for each game in the test set, averaged over up to 1,400 runs, 14 agents, 100 runs per game. Game ticks are displayed on the X axis, maximum 2,000 game ticks. three different predictor models trained on early, mid and late game data features, as well as the baseline rule-based predictor. If F1-scores were 0 for all models, accuracy is plotted instead. Additionally, win average is reported for each game. Games from top to down, left to right: *Dungeon, Eggomania, Escape, Factory Manager, Fireman, Frogs, Ghost Buster* and *Hungry Birds*.

performance in GVGAI. Such system would need to count on an accurate win prediction model (to know if switching to a different method is required) and a second model that determines which is the best method given the features observe (to know what to change to).

Table 4.5 summarizes the F1-scores of the three models on the different game phases identified over all games. Results shown in this table indicate that the rule-based model provides a consistent performance in all game phases. It is better than the others in the early phase (F1-score of 0.42). However, in the middle and late game phases, M_g is significantly better than all the others (F1-scores of 0.57 and 0.71, respectively). Over all games and phases, the middle game M_g model is the best one with an average F1-score of 0.53. M_g is the strongest model in the

Table 4.5: F1-Scores each model per game phase over all games, accuracy in brackets. Each column is a game phase, each row is a model. Highlighted in bold is the best model on each game phase, as well as overall best phase and model.

	Early-P	Mid-P	Late-P	Total-M
Eg	0.22 (0.72)	0.42 (0.74)	0.49 (0.76)	0.38 (0.74)
Mg	0.29 (0.72)	**0.57 (0.79)**	**0.71 (0.83)**	**0.53 (0.78)**
Lg	0.01 (0.73)	0.05 (0.74)	0.22 (0.76)	0.09 (0.74)
Rg	**0.42 (0.67)**	0.47 (0.61)	0.46 (0.58)	0.45 (0.62)
Total-P	0.24 (0.71)	0.38 (0.72)	**0.47 (0.73)**	

individual mid and late phases, only overcome by the simple rule predictor (which incorporates human knowledge as it considers that gaining score leads to a victory) in the early game phase.

Finally, in order to test the robustness of the predictions, we play-tested the test games with MCTS while using the models trained with RHEA. The agent experience features extracted from an MCTS agent are fed into the prediction models trained with a different algorithm. Figure 4.4 shows a comparison between this testing and the previous one. On the left, the model being used by an MCTS agent. On the right, when played with the RHEA and RS variants. As can be seen, all models behave similarly in the different stages of the game and are able to accurately predict the game outcome half-way through it.

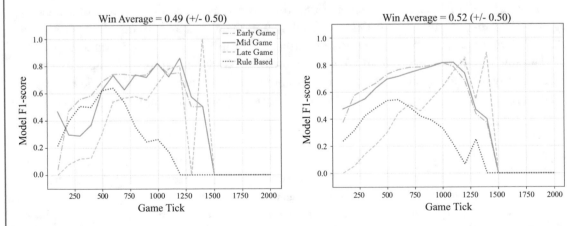

Figure 4.4: F1-scores achieved in the game *Ghost Buster*, trained (in 80 training games) with RHEA and RS and tested (on 20 test games) with MCTS (left) and RHEA (right).

Next Steps

As mentioned before, a logical next step would be to build a hyper-heuristic GVGAI agent that can switch between algorithms, in light of the predictions, when playing any game. Identifying which is the algorithm to switch to can be decomposed in two sub-tasks: one of them signaling which feature measurements need to change and the other identifying which agent can deliver the desired new behavior.

Regarding the first task, it is possible to analyze how the different features influence each task. Figure 4.5 shows an example of the prediction given by the three models at game tick 300 in *Frogs* (level 0, played by 2-8-RHEA). As can be seen, different features indicators are highlighted for each model, signaling which features are *responsible* for the win or loss predictions. It is possible that a hyper-heuristic method that makes uses of this analysis could determine the reasons for the predictions of the models.

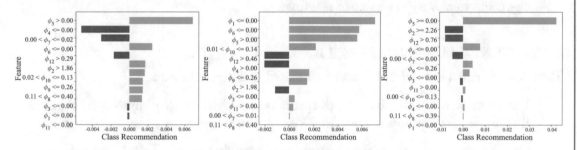

Figure 4.5: Class predictions by features (LIME system[2]). Red signifies the model feature recommends a loss, green a win. The probability of class being selected based on individual feature recommendation is plotted on the X-axis.

Finally, the model could be enhanced with deep variants and more features, such as empowerment [Guckelsberger et al., 2017], spatial entropy or characterization of agent surroundings [Volz et al., 2018].

4.5   EXERCISES

The GVGAI Framework is available in a Github repository.[3] Use the release 2.3[4] in order to run the same version presented in this chapter. Models and data are for the Win Prediction work are also available at a Github repository[5] and you can use a specific checkpoint for the version presented here.[6,7]

---

[2]https://github.com/marcotcr/lime
[3]https://github.com/GAIGResearch/GVGAI
[4]https://github.com/GAIGResearch/GVGAI/releases/tag/2.3
[5]https://github.com/rdgain/ExperimentData/tree/GeneralWinPred-CIG-18/
[6]https://github.com/rdgain/ExperimentData/commit/dc354e6047e378833ef852d0a053aa9215cc6a6b
[7]These exercises are also available at this book's website: https://gaigresearch.github.io/gvgaibook/.

4.5.1   CURRENT PROBLEMS IN GVGAI PLANNING

Section 4.3 presents the main challenges for GVGAI planning at the moment. Interesting projects can arise from trying to tackle the following points.

- Compute game-related features that use an efficient but more accurate measure than Euclidean distance, in order to improve existing methods with them.

- Modify a successful algorithm so it uses a more abstract set of actions, either macro-actions or policies that aim for a higher level objectives (i.e., move the avatar to a specific location, avoid an enemy, collect certain items, etc.).

- Implement an heuristic that is able to switch between different methods depending on certain game-based features. The heuristic should detect and switch to the most appropriate algorithm to play a given game in real time.

4.5.2   GENERAL WIN PREDICTION

The work presented in Section 4.4 can be enhanced in different projects.

- Determine new features to use for the prediction models. Do they improve the algorithms that foresee the outcome of the game? Would they be game- or agent-based features?

- Create a similar system to the one presented here that predicts which algorithm would play better the current game, based on agent-based features only. This new system, in combination with the former one, would be able to establish which algorithm the agent should switch to once the outcome predictions come negative.

- Use different classifiers to train the models. It would be interesting to find out if using deep learning methods provides better predictions than the ones presented here.

CHAPTER 5

Learning in GVGAI

Jialin Liu

The previous chapters mainly discuss the design of the GVGAI framework and planning in GVGAI, in which an FM of each game is available. GVGAI can also be used as a RL platform. In this context, the agent is not allowed to interact with the FM to plan ahead actions to execute in the real game, but need to learn through experience, thus playing the actual game multiple times (as *episodes* in RL), aiming at improving their performance progressively.

Two research questions are conducted. First, can we design a learning agent A_1 which is capable of playing well on unknown levels of game G_1 after having been trained on a few known levels of the same game? Then, on a new game G_2, is it possible to train an agent A_2, based on A_1, more efficiently than from scratch? The first question focus on improving an agent's ability of performing similar tasks with same objectives and rules. An application is the autonomous order picking robots in the warehouses, which can optimize the routes and travel through different picking areas to maximize pick-up efficiency. The second one aims at enhancing learning efficiency on new tasks by transferring the knowledge learned on distinct ones (e.g., tasks with different rules). For instance, a tennis player perhaps masters badminton more quickly than someone who plays neither sport.

This chapter raises first the challenges of learning in GVGAI (Section 5.1). Section 5.2 provides an overall view of the GVGAI learning platform, then present the two environments that have been implemented. Section 5.3 presents the rules and games used in the competitions, the approaches that have been used by the submitted agents to the competitions, together with the analysis of the performance of the different approaches. The competition entries are described in Section 5.4.

5.1 CHALLENGES OF LEARNING IN GVGAI

The planning and learning tasks share some identical challenges including, but not limited to, the lack of *a priori* knowledge, requirement of generality, delayed rewards (usually together with a flat reward landscape) and real-time constraints.

Lack of *a priori* knowledge   The lack of *a priori* knowledge is two-fold. First, game rules are not provided. The agent objective is to win a game but it has no knowledge about how to score, win/lose or any of the termination conditions. Only the agent's state observation at current game tick, including the legal actions in the current state, is accessible. However, this state observation does not necessarily cover the whole game state. Second, in the corresponding competitions, only some of the game levels are released for training; the agents will be tested on unseen levels.

Real-time constraints   Due to the real-time feature of video games, the construction and initialization of a learning agent should not take too long, and during the game playing, an agent needs to decide an action rapidly enough (in the GVGAI competitions, no more than 1 s and 40 ms, respectively). This ensures a smoother playing experience, similar to when a human player plays a real-time video game. For this reason, in the competitions presented later in Section 5.3, a learning agent is disqualified in the competition or a *doNothing* action is performed instead of its selection depending on the actual time it takes to select an action.

General game playing   Another issue is that an agent trained to perform well on a specific level of a specific game may perform fairly on a similar game level but fails easily on a very different game or level. For instance, a human-level *Super Mario* agent may not survive long in a *PacMan* game; an agent which dominates in *Space Invaders* is probably not able to solve *Sokoban*. Designing a single agent that performs well on a set of different unknown games is not trivial.

Delayed rewards   In all the games in the GVGAI framework (as well as in the competitions), the main task is to win the game. However, no winning condition is provided but the instant game score at the current episode. In puzzle and board games, the landscape of instant scores during game-play is usually flat until the last episode, therefore the design of some heuristic and planning ahead are necessary. Defining an appropriate heuristic is not trivial neither for unseen levels of a game or for a set of games.

5.2   FRAMEWORK

Different from the planning tracks, no FM is given to the agents in the learning environment, thus, no simulation of games is available. To avoid accessing to the forward model, a new interface was implemented in 2017 on top of the main GVGAI framework. Then, *Philip Bontrager* and *Ruben Rodriguez Torrado* interfaced it with OpenAI Gym in 2018. The former one has been used in the first GVGAI Learning Competition organized at the IEEE's 2017 Conference on Computational Intelligence and Games (IEEE CIG 2017), which supports agents written in *Java* and *Python*. Then the second competition was organized at the IEEE's 2018 CIG using a modified framework interfaced with OpenAI Gym, which only supports agents written in *Python*. The GVGAI framework makes it easy to train and test learning agents potentially on an infinity number of games thanks to the use of VGDL. This is a difficult RL problem due to the required generality and limited online decision time.

5.2.1   GVGAI LEARNING FRAMEWORK

A number of games are provided in each game set of which each includes a number of levels (usually five in the framework). Thanks to VGDL and the integrated games provided by the GVGAI framework, this enables the option to design different sets of games and levels.

Execution in a set would have two phases: a *Learning Phase* using N_L levels of the N available and a *Test Phase*, using the other N_T levels with $N_T = N - N_L$. The big picture of these two phases are given below, though there are some differences in details in the 2017's and 2018's competitions.

Learning phase. An agent has a limited amount of time, T_L, for playing a set of M games with N_L learning levels each. It will play each of the N_L levels once, then to be free to choose the next level to play if there is time remaining. The agent is allowed to send the action abort to finish the game at any moment, apart from the normal available game actions. The method result is called at the end of every game playing regardless if the game has finished normally or been corrupted by the agent using abort. Thus, the agent can play as many games as desired, potentially choosing the levels to play in, as long as it respects the time limit T_L.

Test phase. After the learning phase, the trained agent is tested on the N_T test levels that have never been seen during learning. The win rate and statistics of the score and game length over the test trials are used to evaluate the agents.

Note that no FM is accessible during neither of the phases, but learning agents receive an observation of the current game state at every game tick in different form(s), depending on which GVGAI learning environment is used. Both environments are available on-line and can be used for research on general video game learning.

5.2.2   GVGAI LEARNING ENVIRONMENT

The GVGAI Learning Environment used in the IEEE CIG 2017's Single-Player Learning Competition is briefly introduced in this section, more technical details about how to set up the framework and the competition procedure are described in the technical manual [Liu, Pérez Liébana, and Lucas, 2017b].

The GVGAI Learning Environment is part of the same project as the planning environment.[1] It supports learning agents written in *Java* or *Python*. At every game tick, the agent is allowed to send a valid game action to execute in the game, and, at the same time, to select the format of the next state observation to receive between a serialized JSON game observation (*String*) and a screenshot of the game screen (*PNG*). At any time, the agent can select the format of the game observation to be received at next game tick, using one of the following types:

```
lastSsoType = Types.LEARNING_SSO_TYPE.JSON; //request for a JSON
lastSsoType = Types.LEARNING_SSO_TYPE.IMAGE; //request for a
```

---

[1]https://github.com/GAIGResearch/GVGAI

```
screen-shot
lastSsoType = Types.LEARNING_SSO_TYPE.BOTH; //request for both
```

The choice will be remembered until the agent makes another choice using the above commands. An example of the screenshot is given in Figure 5.1.

Figure 5.1: Example: screenshot of a game screen.

Below is an example of the serialized state observation.

```
SerializableStateObservation phase=ACT, isValidation=false,
gameScore=8.0, gameTick=150, gameWinner=NO_WINNER, isGameOver=false,
worldDimension=[230.0, 200.0], blockSize=10, noOfPlayers=1,
...
availableActions=[ACTION_USE, ACTION_LEFT, ACTION_RIGHT],
avatarResources={}, observationGrid={ Observation{category=6, itype=0,
obsID=578, position=0.0 : 0.0, reference=-1.0 : -1.0, sqDist=2.0}
...
Observation{category=6, itype=11, obsID=733, position=150.0 : 40.0,
reference=-1.0 : -1.0, sqDist=24482.0} }, resourcesPositions=null,
portalsPositions=null, fromAvatarSpritesPositions=null}
```

As in the planning environment, a *Java* or *Python* agent should inherit from an abstract class AbstractPlayer, implement the constructor and three methods: act, init and result. The class must be named Agent.java or Agent.py.

Implementation of a Learning Agent

`Agent()`   The constructor is called once per game and must finish in no more than `START_TIME` (by default 1 s) of CPU time, thus it is called once during the learning and test phases of each game.

`init(SerializableStateObservation sso, ElapsedCpuTimer elapsedTimer)`   After creating an agent, `init()` is called before every single game run. It should finish in no more than `INITIALIZATION_TIME` (by default 1 s) of CPU time.

`act(SerializableStateObservation sso, ElapsedCpuTimer elapsedTimer)`   At each game tick, `act()` is called and determines the next action of the agent within the prescribed CPU time `ACTION_TIME` (by default 40 ms). The possible actions are `ACTION_LEFT`, `ACTION_RIGHT`, `ACTION_UP`, `ACTION_DOWN`, `ACTION_USE` and `ACTION_NIL` (do nothing). The agent will be disqualified immediately if more than `ACTION_TIME_DISQ` (by default 50 ms) is taken. Otherwise, a `NIL` action (do nothing) is applied. Note that it is possible that in a game or at a game state, not all the actions listed above are available (legal) actions, but `ACTION_NIL` always is.

`result(SerializableStateObservation sso, ElapsedCpuTimer elapsedTimer)`
This method is called at the end of every game. It has no time limit, so the agent doesn't get penalized for overspending other than the `TOTAL_LEARNING_TIME` indicated. The agent can play with the time it spends on the `result` call to do more learning or to play more games. At each call of `result`, an action or a level number should be returned.

Termination   A game playing terminates when the player wins/loses the game or the maximal game ticks (`MAX_TIMESTEPS`) is reached.

Time out   If the agent returns an action after `ACTION_TIME` but no more than `ACTION_TIME_DISQ`, then the action `ACTION_NIL` will be performed.

Disqualification   If the agent returns an action after `ACTION_TIME_DISQ`, the agent is disqualified and loses the game.

Parameters
The notation and corresponding parameters in the framework are summarized in Table 5.1, as well as the default values.

5.2.3   GVGAI GYM ENVIRONMENT

GVGAI Gym is a result of interfacing the GVGAI framework to the OpenAI Gym environment by Ruben Rodriguez Torrado and Philip Bontrager, Ph.D. candidates at the New York University School of Engineering [Torrado et al., 2018]. Beside the more user-friendly interface, a learning agent still receives a screenshot of the current game screen and game score, then returns a valid action at every game tick. An example of a random agent that plays first

Table 5.1: The main parameters in the learning framework

Parameters for Client (Agent)		
Variable	Default Value	Usage
START_TIME	1 s	Time for agent's constructor
initialization_TIME	1 s	Time for init()
ACTION_TIME	40 ms	Time for returning an action per tick
ACTION_TIME_DISQ	50 ms	Threshold for disqualification per tick
TOTAL_LEARNING_TIME	5 min	Time allowed for learning a game
EXTRA_LEARNING_TIME	1 s	Extra learning time
SOCKET_PORT	8080	Socket port for communication
Parameters for Client Server		
Variable	Default	Usage
MAX_TIMESTEPS	1000	Maximal game ticks a game can run
VALIDATION_TIMES	10	Number of episodes for validation

```
1  import gym
2  import gym_gvgai
3
4  env = gym.make('gvgai-aliens-lvl0-v0')
5  env.reset()
6
7  score = 0
8  for i in range(2000):
9      action_id = env.action_space.sample()
10     state, reward, isOver, info = env.step(action_id)
11     score += reward
12     print("Action " + str(action_id) + " played at game tick " + str(i+1) + ", reward=" + str(reward) + ", new score=" + str(score))
13     if isOver:
14         print("Game over at game tick " + str(i+1) + " with player " + info['winner'])
15         break
```

Figure 5.2: Sample code of randomly playing the first level of *Aliens* using GVGAI Gym.

level of *Aliens* using GVGAI Gym is illustrated in Figure 5.2. We compare the GVGAI Gym implementation and the original GVGAI environment in Table 5.2.

5.2.4   COMPARING TO OTHER LEARNING FRAMEWORKS

Other general frameworks like OpenAI Gym [Brockman et al., 2016], Arcade Learning Environment (ALE) [Bellemare et al., 2013] or Microsoft Malmö [Johnson et al., 2016] contain a great number of single-/multi-player, model-free or model-based tasks. Interfacing with these systems would greatly increase the number of available games which all GVGAI agents could play via a common API. This would also open the framework to 3D games, an important section of the environments the current benchmark does not cover.

Table 5.2: Comparison of the planning and learning environments

	GVGAI Planning		GVGAI Learning	GVGAI Gym
	1-Player	2-Player		
Similarities	• Play unseen games, no game rules available • Access to game score, tick, if terminated • Access to legal actions • Access to observation of current game state			
Forward model?	Yes		No	NO
History events?	Yes		No	NO
State Observation?	*Java* object		*String* or PNG	*PNG*
	Java		Java & Python	Python

At the time of writing, ALE [Bellemare et al., 2013] offers higher-quality games than GVGAI as they were home-console commercial games of a few decades ago, whereas the GV-GAI provides a structured API (information available via *Java* objects, or *JSON* interface, or screen capture); the agents are tested on unseen games; and there is potentially infinite supply of games. In GVGAI terms, ALE offers just two tracks: single-player learning and planning, with the learning track being the more widely used. The GVGAI framework has the potential to be expanded by adding a two-player learning track, which will offer more open-ended challenges. This is outside of the current scope of ALE. Again, thanks to VGDL, it is much more easier to create new games or to create new levels for these games, using the GVGAI Learning environment or GVGAI Gym. It is also easy to automatically generate variations on existing VGDL games and their levels. Thus, the users can apply procedural content generation to generate game and level variations for training and testing their learning agents. GVGAI is more easily extensible than ALE, and offers a solution to overfitting.

5.3 GVGAI LEARNING COMPETITIONS

At the time that this book is written, only two GVGAI learning competitions have been organized. In this section, we describe the competition rules and core challenges of individual competition besides the common challenges of both that have been presented in Section 5.1.

5.3.1 COMPETITION USING THE GVGAI LEARNING ENVIRONMENT

The first GVGAI learning competition was organized at the IEEE's 2017 Conference on Computational Intelligence and Games (IEEE CIG 2017).

Competition Procedure and Rules

In this competition, 10 games are used, of which 3 levels are given for training and 2 private levels are used for testing. The set of 10 games used in this learning competition is the training set 1 of the 2017 GVGAI Single-Player Planning Competition.

The *Learning Phase* consists of two steps, referred to as *Learning Phase 1* and *Learning Phase 2*. The whole procedure is illustrated in Algorithm 5.6 and Figure 5.3. An agent has a limited duration 5 min (legal learning duration) in total for both training phases. The communication time is not included by the Timer. In case that 5 min has been used up, the results and observation of the game will still be sent to the agent and the agent will have no more than 1 s before the test.

Algorithm 5.6 Main procedure of the 2017 learning competition.

Require: $\mathcal{G}$ set of games
Require: $\mathcal{L}$ set of training levels (per game)
Require: $\mathcal{T}$ set of training levels (per game)
Require: π and agent

 1: **for** each game $G \in \mathcal{G}$ **do**
 2:    **for** each game *level* $\in G_{\mathcal{L}}$ **do**
 3:       Let π play *level* once
 4:       *nextLevel* $\leftarrow \pi$.result()
 5:    **end for**
 6:    **while** Time is not elapsed **do**
 7:       Let π play *nextLevel* once
 8:       *nextLevel* $\leftarrow \pi$.result()
 9:    **end while**
10: **end for**
11: **for** each game $G \in \mathcal{G}$ **do**
12:    **for** each game *level* $\in G_{\mathcal{T}}$ **do**
13:       $\text{RES}_G \leftarrow$ Results of 10 games of π on *level*
14:    **end for**
15: **end for**
16: **Return** $\{\text{RES}_G\}$

During the *Learning phase 1* of each game (lines 2–5 of Algorithm 5.6), an agent plays once the three training levels sequentially. At the end of each level, whether the game has terminated normally or the agent forces to terminate the game (using abort), the server will send the results of the (possibly unfinished) game to the agent before termination.

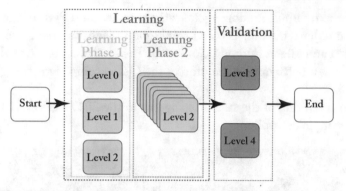

Figure 5.3: Learning and test phases for one game in the 2017 single-player learning competition. In the competition, an agent will be executed on a set of (usually 10) unknown games, thus this process will repeat 10 times.

After having finished *Learning phase 1*, the agent is free to select the next level to play (from the three training levels) by calling the method `int result()` (detailed in Section 5.2.2). If the returned index of the selected level is not a valid index, then a random index from the valid indexes will be passed and a new game will start. This step (*Learning Phase 2*) (lines 6–9 of Algorithm 5.6) is repeated until the legal learning time has expired.

After looping over the whole set of games, the *Test Phase* (lines 11–15 of Algorithm 5.6) starts. The trained agent repeatedly plays 10 times the private test levels sequentially. There is no more total time limit, but the agent still needs to respect the time limits for the methods `init`, `act` and `result`, and can continue learning during the game playing.

Challenges of the Competition

Beside the challenges of learning in GVGAI (Section 5.1), there are some other crucial problems due to the competition rules, such as how to select which level to train next; how to distribute the total learning duration; which type of state observation to receive; training different models for different games or training one unique model; etc. None of these is trivial to decide and no entry performed reasonably in the competition. Therefore, the competition rules were changed for 2018.

Competition Results

In the 2017 edition of GVGAI learning competition, the execution of controllers was divided into two phases: learning and validation. In the learning phase, each controller has a limited amount of time, 5 min, for learning the first three levels of each game. The agent could play as many times as desired, choosing among these three levels, as long as the 5 min time limit is respected. In the validation phase, the controller plays 10 times the levels 4 and 5 sequentially. The results obtained in these validation levels are the ones used in the competition to rank the entries. Besides the two sample random agents written in Java and Python and one sample agent using Sarsa written in Java, the first GVGAI single-player learning track received three submissions written in Java and one in Python [Liu, 2017]. The results are illustrated in Table 5.3. The winner of this track is a naive implementation of the Q-Learning algorithm (Section 5.4.5).

Table 5.3: Score and ranking of the submitted agents in the 2017's GVGAI Learning Competition. †denotes a sample controller.

Agent	Training Set		Test Set	
	Score	Ranking	Score	Ranking
kkunan	125	6	184	1
sampleRandom†	154	2	178	2
DontUnderestimateUchiha	149	3	158	3
sampleLearner†	149	4	152	4
ercumentilhan	179	1	134	5
YOLOBOT	132	5	112	6

Table 5.4 compares the best scores by single-player planning and learning agents on the same test set. Note that one of the games in the test set is removed from the final ranking due to bugs found in the game itself. The best performance of learning agents on tested games is far worse than what the planning agents can achieve.

5.3.2   COMPETITION USING THE GVGAI GYM

The GVGAI Gym has been used in the learning competition organized at the IEEE's 2018 Conference on Computational Intelligence and Games (IEEE CIG 2018).

Competition Procedure and Rules
The competition rules have been changed based on the experience of the first GVGAI learning competition. The main procedures are as follows.

1. Three games with two public levels each were given for training one month before the submission deadline. The participants were free to train their agents privately with no constraints and can use any computational resource that they had.

2. After the submission deadline, three new private levels of each of the three games are used for validation. The validation phase was ran by the organizers using one single laptop. Each experiment has been repeated 10 times. During validation, each agent has 100 ms per game tick to select an action.

Table 5.4: Table compares the best scores by single-player planning and learning agents on the same test set. Note that one of the games in the test set is removed from the final ranking due to bugs in the game itself. †denotes a sample controller.

Game	1-P Planning	1-P Learning	
	Best Score	Best Score	Agent
G2	109.00 ± 38.19	31.5 ±14.65	sampleRandom†
G3	1.00 ± 0.00	0 ± 0	*
G4	1.00 ± 0.00	0.2 ± 0.09	kkunan
G5	216.00 ± 24.00	1 ± 0	*
G6	5.60 ± 0.78	3.45 ± 0.44	DontUnderestimateUchiha
G7	31,696.10 ± 6,975.78	29,371.95 ± 2,296.91	kkunan
G8	1,116.90 ± 660.84	35.15 ± 8.48	kkunan
G9	1.00 ± 0.00	0.05 ± 0.05	sampleRandom†
G10	56.70 ± 25.23	2.75 ± 2.04	sampleLearner†

Three test games with very different aspects have been designed for this competition. The game 1 is modified from *Aliens*, and the only difference is that the avatar is allowed to shoot and move in four directions instead of one and two directions, respectively. The game 2 is a puzzle game called *Lights On*, in which the avatar wins if it turns on the lights shown on the right end of the game screen (e.g., Figure 5.4). The game 3 is modified from a deceptive game by Anderson et al. [2018], called *DeceptiCoins*.

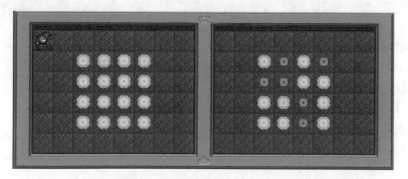

Figure 5.4: Screenshot of the puzzle game *Lights On* that has been used in the second GVGAI learning competition. The screenshot illustrates the initial state of a level.

Challenges of the Competition

The new competition rules and more user-friendly framework make the users focus more on the learning algorithms, including those that are already compatible with Open AI Gym. The fact that users are free to train their agents privately using as much as learning time and computational resources as they like enables more opportunities and possibilities. However, some classic learning algorithms are not able to handle the different dimensions of the screen that distinct games have.

Competition Results

This edition of the competition received only 2 entries, *fraBot-RL-QLearning* and *fraBot-RL-Sarsa*, submitted by the same group of contributors from the Frankfurt University of Applied Science. The results of the entries and sample agents (*random, DQN, Prioritized Dueling DQN* and *A2C* [Torrado et al., 2018]) are summarized in Table 5.5. For comparison, the planning agent *OLETS* (with access to the forward model) is included. *DQN* and *Prioritized Dueling DQN* are outstanding on level 3 (test level) of the game 1, because the level 3 is very similar to the level 2 (training level). Interestingly, the sample learning agent *DQN* outperformed *OLETS* on the third level of game 1. For instance, the baseline agents *DQN, Prioritized Dueling DQN* and *A2C* are not able to learn the game *DeceptiCoins* due to the different game screen dimensions in different levels, despite their outstanding performance on the test levels of the game modified from *Aliens*.

Table 5.5: Score and ranking of the submitted agents in the 2018's GVGAI Learning Competition. [†]denotes a sample controller.

| Game | Game 1 | | | Game 2 | | | Game 3 | | | Ranking |
Level	3	4	5	3	4	5	3	4	5	
fraBot-RL-Sarsa	-2	1	-1	0	0	0	2	3	2	1
fraBot-RL-QLearning	-2	-1	-2	0	0	0	1	0	2	2
Random[††]	-0.5	0.2	-0.1	0	0	0	3.5	0.7	2.7	3
DQN[†]	61.5	-1	0.3	0	0	0	-	-	-	-
Prioritized Dueling DQN[†]	36.8	-1	-2	0	0	0	-	-	-	-
A2C[†]	8.1	-1	-2	0	0	0	-	-	-	-
OLETS Planning Agent	41.7	48.6	3.1	0	0	2.2	4.2	8.1	14	-

5.4   COMPETITION ENTRIES

This section first describes the approaches that tackled the challenge set in the single-player learning track of the 2017 and 2018 competitions, and then moves to other approaches.

5.4.1   RANDOM AGENT (SAMPLE AGENTS)

A sample random agent, which selects an action uniformly at random at every game tick, is included in the framework (in both Java and Python) for the purposes of testing. This agent is also meant to be taken as a baseline: a learner is expected to perform better than an agent which acts randomly and does not undertake any learning.

5.4.2   DRL ALGORITHMS (SAMPLE AGENTS)

Using the new GVGAI Gym, Torrado et al. [2018] compared three implemented Deep Reinforcement Learning algorithms of the OpenAI Gym, Deep Q-Network (DQN), Prioritized Dueling DQN, and Advance Actor-Critic (A2C), on eight GVGAI games with various difficulties and game rules. All the three RL agents perform well on most of the games, however, DQNs and A2C perform badly when no game score is given during a game playing (only win or loss is given when a game terminates). These three agents have been used as sample agents in the learning competition organized at IEEE CIG 2018.

5.4.3   MULTI-ARMED BANDIT ALGORITHM

DontUnderestimateUchiha by K. Kunanusont is based on two popular multi-armed bandit (MAB) algorithms; ϵ-Decreasing Greedy Algorithm and upper confidence bounds (UCB). At any game tick T, the current *best* action with probability $1 - \epsilon_T$ is picked, otherwise an action is uniformly randomly selected. The *best* action at time T is determined using UCB with increment of score as reward. This is a very interesting combination, as the UCB-style selection and the ϵ-Decreasing Greedy Algorithm both aim at balancing the trade-off between exploiting more the best-so-far action and exploring others. Additionally, ϵ_0 is set to 0.5 and it decreases slowly along time, formalized as $\epsilon_T = \epsilon_0 - 0.0001T$. According to the competition setting, all games will last longer than 2,000 game ticks, so $\forall T \in \{1, \ldots, 2000\}, 0.5 \geq \epsilon_T \geq 0.3$. As a result, random decisions are made for approximately 40% time.

5.4.4   SARSA

sampleLearner, *ercumentilhan* and *fraBot-RL-Sarsa* are based on the State-Action-Reward-State-Action (Sarsa) algorithm [Russell and Norvig, 2016]. The *sampleLearner* and *ercumentilhan* use a subset of the whole game state information to build a new state to reduce the amount of information to be saved and to take into account similar situations. The main difference is that the former uses a square region with fixed size centered at the avatar's position, while the latter uses a first-person view with a fixed distance. *fraBot-RL-Sarsa* uses Sarsa, and it uses the entire screenshot of the game screen as input provided by GVGAI Gym. The agent has been trained using 1,000 episodes for each level of each game, and the total training time was 48 h.

5.4.5 Q-LEARNING

kkunan, by K. Kunanusont, is a simple Q-learning [Russell and Norvig, 2016] agent using most of the avatar's current information as features, which a few exceptions (such as avatar's health and screen size, as these elements that vary greatly from game to game). The reward at game tick $t + 1$ is defined as the difference between the score at $t + 1$ and the one at t. The learning rate α and discounted factor γ are manually set to 0.05 and 0.8. During the *learning phase*, a random action is performed with probability $\epsilon = 0.1$, otherwise, the best action is selected. During the *validation phase*, the best action is always selected. Despite it's simplicity, it won the first track in 2017. *fraBot-RL-QLearning* uses the Q-Learning algorithm. It has been trained using 1,000 episodes for each level of each game, and the total training time was 48 hours.

5.4.6 TREE SEARCH METHODS

YOLOBOT is an adaption of the *YOLOBOT* planning agent (as described previously in Chapter 4). As the FM is no more accessible in the learning track, the MCTS is substituted by a greedy algorithm to pick the action that minimizes the distance to the chosen object at most. According to the authors, the poor performance of *YOLOBOT* in the learning track, contrary to its success in the planning tracks, was due to the collision model created by themselves that did not work well.

5.4.7 OTHER LEARNING AGENTS

One of the first works that used this framework as a learning environment was carried out by Samothrakis et al. [2015], who employed Neuro-Evolution in 10 games of the benchmark. Concretely, the authors experimented with separable natural evolution strategies (S-NES) using two different policies (ϵ-greedy vs. Softmax) and a linear function approximator vs. a neural network as a state evaluation function. Features like score, game status, avatar and other sprites information were used to evolve learners during 1,000 episodes. Results show that ϵ-greedy with a linear function approximator was the better combination to learn how to maximize scores on each game.

Braylan and Miikkulainen [2016] performed a study in which the objective was to learn a forward model on 30 games. The objective was to learn the next state from the current one plus an action, where the state is defined as a collection of attribute values of the sprites (spawns, directions, movements, etc.), by means of logistic regression. Additionally, the authors transfer the learned object models from game to game, under the assumption that many mechanics and behaviors are transferable between them. Experiments showed the effective value of object model transfer in the accuracy of learning forward models, resulting in these agents being stronger at exploration.

Also in a learning setting, Kunanusont [2016] and Kunanusont, Lucas, and Pérez Liébana [2017] developed agents that were able to play several games via screen capture. In particular, the authors employed a Deep Q-Network in seven games of the framework of increasing complexity, and included several enhancements to GVGAI to deal with different screen sizes and a non-visualization game mode. Results showed that the approach allowed the agent to learn how to play in both deterministic and stochastic games, achieving a higher winning rate and game score as the number of episodes increased.

Apeldoorn and Kern-Isberner [2017] proposed a learning agent which rapidly determines and exploits heuristics in an unknown environment by using a hybrid symbolic/sub-symbolic agent model. The proposed agent-based model learned the weighted state-action pairs using a sub-symbolic learning approach. The proposed agent has been tested on a single-player stochastic game, *Camel Race*, from the GVGAI framework, and won more than half of the games in different levels within the first 100 game ticks, while the standard Q-Learning agent never won given the same game length. Based on Apeldoorn and Kern-Isberner [2017], Dockhorn and Apeldoorn [2018] used exception-tolerant hierarchical knowledge bases (HKBs) to learn the approximated forward model and tested the approach on the 2017 GVGAI Learning track framework, respecting the competition rules. The proposed agent beats the best entry in the learning competition organized at CIG 2017 [Dockhorn and Apeldoorn, 2018], but still performed far worse than the best planning agents, which have access to the real forward models.

Finally, Justesen et al. [2018] implemented A2C within the GVGAI-Gym interface in a training environment that allows learning by procedurally generating new levels. By varying the levels in which the agent plays, the resulting learning is more general and does not overfit to specific levels. The level generator creates levels at each episode, producing them in a slowly increasing level of difficulty in response to the observed agent performance.

5.4.8 DISCUSSION

The presented agents differ between each other in the input game state (JSON string or screen capture), the amount of learning time, the algorithm used. Additionally, some of the agents have been tested on a different set of games and sometimes using different game length (i.e., maximal number of game ticks allowed). None of the agents, which were submitted to the 2017 learning competition, using the classic GVGAI framework, have used screen capture.

The Sarsa-based agents performed surprisingly bad in the competition, probably due to the arbitrarily chosen parameters and very short learning time. Also, learning three levels and testing on two more difficult levels given only 5 min learning time is a difficult task. An agent should take care of the learning budget distribution and decide when to stop learning a level and to proceed the next one.

The learning agent using exception-tolerant HKBs [Dockhorn and Apeldoorn, 2018] learns fast. However, when longer learning time is allowed, it is dominated by deep reinforcement learning (DRL) agents. Out of the eight games tested by Torrado et al. [2018], none of the tested three DRL algorithms outperformed the planning agents on six games. However, on the heavily stochastic game Seaquest, A2C achieved almost double score than the best planning agent, MCTS.

5.5 SUMMARY

In this chapter, we present two platforms for the GVGAI learning challenges, which can be used for testing reinforcement learning algorithms, as well as some baseline agents. This chapter also reviews the 2017 and 2018 GVGAI learning competitions organized using each of the platforms. Thanks to the use of VGDL, the platforms have the potential of designing new games by humans and AIs for training reinforcement learning agents. In particular, the GVGAI Gym is easy to use to implement and compare agents. We believe this platform can be used in multiple research directions, including designing reinforcement learning agents for a specific task, investigating artificial general intelligence, and evaluating how different algorithms can learn and evolve to understand various changing environments.

5.6 EXERCISES

These exercises are also available at this book's website: https://gaigresearch.github.io/gvgaibook/.

5.6.1 DOWNLOAD AND INSTALLATION

All the code of GVGAI learning environment is available in a GitHub repository.[2] Use the checkpoint[3] in order to run the same version presented in this chapter. All the code of GVGAI Gym is available in a GitHub repository,[4] as well as the instructions. Use the checkpoint[5] in order to run the same version presented in this chapter. We recommend to use the GVGAI Gym implementation. A step-by-step tutorial video about how to download, install and run the code using Jupyter notebook can be found on YouTube.[6]

5.6.2 PLAY A GAME RANDOMLY

The exercise is to randomly play a level of a game, for example *Aliens*. We recommend doing this using Jupyter notebook. This enables the code to be easily modified and re-run. The sample code is given in Figure 5.2. You can ran the sample code directly. The line below indicates the environment to use.

```
env = gym.make(`gvgai-aliens-lvl0-v0')
```

If you would like to play some other games, please replace "aliens" with another one (i.e., "escape" or "bait"). A full list of available games can be found in the folder **gym_gvgai/envs/games**. For

---

[2]https://github.com/GAIGResearch/GVGAI
[3]https://github.com/GAIGResearch/GVGAI/commit/a0e267416f27b5a0ac5b30644ed0354a6e4b4705
[4]https://github.com/rubenrtorrado/GVGAI_GYM
[5]https://github.com/rubenrtorrado/GVGAI_GYM/commit/20cac6e9e783e3ef381eb87446a21f6bb21557e3
[6]https://www.youtube.com/watch?v=O84KgRt6AJI&feature=youtu.be

each of the games, there are normally five levels. You can indicate the level that you want to play by replacing "lvl0." For instance, you can play the second level of game *Zelda* using `env = gym.make(`gvgai-zelda-lvl1-v0')`.

5.6.3 TRAIN A MORE SOPHISTICATED AGENT

The exercise is to train a more sophisticated agent, e.g., A2C. Implementations of some classic reinforcement learning algorithms are available in the GitHub repository, *OpenAI Baselines*.[7] Otherwise, you can try *Stable Baselines*,[8] the improved implementations based on *OpenAI Baselines*.

5.6.4 CREATE YOUR OWN AGENTS

A recent survey [Justesen et al., 2017] showcases a myriad of deep learning methods for games that can be used in GVGAI. The generality and flexibility of this benchmark allows for experimental setup that can go beyond the current state of the art. Examples are as follows.

- Train your method in multiple levels at the same time, to make sure the algorithm learns concepts beyond particular levels. Test them in other levels of the game.

- Modify the VGDL description of a given game to create slight variations of the same game. Could your algorithm learn faster (or better) by training in a set of games that are closer in the game space?

- Train your algorithm in different games and test them in some others. Maybe you can group them by genre first and study the learning capabilities of these methods within the same type of game.

---

[7]https://github.com/openai/baselines/
[8]https://github.com/hill-a/stable-baselines

CHAPTER 6

Procedural Content Generation in GVGAI

Ahmed Khalifa and Julian Togelius

Procedural content generation (PCG) is to use a computer program/algorithm to generate game content [Shaker, Togelius, and Nelson, 2016] automatically. This content could be anything in the game such as textures [Turk, 1991], levels [Khalifa et al., 2016, Nichols, 2016], rules [Khalifa et al., 2017], etc. PCG might be a little new in research but it has been around since the beginning of computer games. It was used as a game design element to allow replayability and new scenarios such as in Rogue (Glenn Wichman, 1980). Also, it was used to allow new types of games with huge amount of content using a small foot print such as in Elite (David Braben and Ian Bell, 1984). Although more and more people have been adapting and using PCG in their games and discovering new techniques, no one was tackling the problem of generality. Most of the used techniques in the industry are all constructive [Shaker et al., 2016] and depend on a lot of hacks and tricks based on the game knowledge. Algorithms that can adapt and work between different games are still a dream for the PCG field. That is the main motivation behind having a PCG track in the GVGAI. We wanted to give the people a framework to help them research toward finding more generic ways that can generate new content with small amount of information about the current game.

GVGAI framework has several different facets that define games: levels, interaction sets, termination conditions, and sprite sets (graphical representation and types). We decided to start with tackling the level generation problem first as level generation is one of the oldest and challenging problems that people are tackling since the early 1980s. Later, we designed a new track for rule generation where the user has to generate both interaction sets and termination conditions given the rest of the facets. One of the core challenges for both tracks is how to define a good way to compare the different generation techniques. For the competition, we relied on humans to provide us with this data based by comparing two generate content from different generators. The humans do the comparisons several times on examples of each generator from different games, to make sure that the algorithm can adapt between different games and also can generate different levels for the same game. This technique works fine for the competition, but it doesn't help the users to understand how to enhance their generator, as there is no specific set of rules that define what a good level or game is. In this chapter, we will talk about these two different tracks: level (Section 6.1) and rule generation (Section 6.2). For each one of them, we will discuss their interface, the sample generators provided, and the latest techniques and generators used.

6.1   LEVEL GENERATION IN GVGAI

The level generation track was introduced in 2016 as a new competition track [Khalifa et al., 2016]. It is considered the first content generation track for the GVGAI framework. The competition focuses on creating playable levels providing the game description. In this track, every participant submits a level generator that produces a level in a fixed amount of time. The framework provides the generator with the game sprites, interaction set, termination conditions, and level mapping, in return, the generator produces a 2D matrix of characters, where each character represents the game sprites at that location. The framework also allows the generator to provide their own level mapping if needed.

The game information is provided to the generator using a `GameDescription` object and `ElpasedCpuTime` object. Figure 6.1 shows the `AbstractLevelGenerator` class that the participants should extend to create their level generator. A `GameDescription` object is a structured data that provide access to the game information in a more organized manner, while the `ElpasedCpuTime` object provides the user with the remaining time for the generator to finish. When running the competition, the `ElpasedCpuTime` timer gives each generator five hours to finish its job.

Beside this object, the framework provides a helper class called `GameAnalyzer` which analyzes the `GameDescription` object and provide additional information that can be used by the generator. The `GameAnalyzer` divides the game sprites into five different types.

- Avatar Sprites: sprites that are controlled by the player.

- Solid Sprites: static sprites that prevent the player movement and have no other interactions.

- Harmful Sprites: sprites that can kill the player or spawn another sprite that can kill the player.

```
public abstract class AbstractLevelGenerator {

    public abstract String generateLevel(GameDescription game, ElapsedCpuTimer elapsedTimer);

    public HashMap<Character, ArrayList<String>> getLevelMapping()
    {
        return null;
    }
}
```

Figure 6.1: AbstractLevelGenerator class functions.

- Collectible sprites: sprites that the player can destroy upon collision with them, providing score for the player.

- Other sprites: any other sprites that doesn't fall in the previous categories.

The `GameAnalyzer` also provides two additional arrays for spawned sprites and goal sprites. It also provides a priority value for each sprite based on the number of times each sprite appears either in interactions or termination conditions. The spawned sprites array contains all the sprites that can be generated from another sprite, while the goal sprites array contains all the sprites that appear in the termination conditions of the game. Table 6.1 shows the `GameAnalyzer` output of the VGDL game *Zelda* defined in Listing 6.1. `nokey`, `withkey`, and `goal` are the only goal

Table 6.1: The *GameAnalyzer* data for the game of *Zelda*

Sprite	Sprite Image	Sprite Type	Spawned Sprite	Goal Sprite	Priority Value
floor		Other	FALSE	FALSE	0
goal		Other	FALSE	TRUE	3
key		Collectible	FALSE	FALSE	1
sword		Other	TRUE	FALSE	1
nokey		Avatar	FALSE	TRUE	4
withkey		Avatar	FALSE	TRUE	2
monsterQuick		Harmful	FALSE	FALSE	4
monsterNormal		Harmful	FALSE	FALSE	4
monsterSlow		Harmful	FALSE	FALSE	4
wall		Solid	FALSE	FALSE	1

```
1  BasicGame
2   SpriteSet
3     floor    > Immovable randomtiling=0.9 img=oryx/floor3 hidden=True
4     goal     > Door img=oryx/doorclosed1
5     key      > Immovable img=oryx/key2
6     sword    > OrientedFlicker limit=5 singleton=True img=oryx/slash1
7     movable >
8       avatar   > ShootAvatar stype=sword frameRate=8
9         nokey        > img=oryx/swordman1
10        withkey      > img=oryx/swordmankey1
11      enemy    > RandomNPC
12        monsterQuick  > cooldown=2 cons=6  img=oryx/bat1
13        monsterNormal > cooldown=4 cons=8  img=oryx/spider2
14        monsterSlow   > cooldown=8 cons=12 img=oryx/scorpion1
15    wall     > Immovable autotiling=true img=oryx/wall3
16
17  InteractionSet
18    movable wall     > stepBack
19    nokey   goal     > stepBack
20    goal    withkey  > killSprite scoreChange=1
21    enemy   sword    > killSprite scoreChange=2
22    enemy   enemy    > stepBack
23    avatar  enemy    > killSprite scoreChange=-1
24    nokey   key      > transformTo stype=withkey scoreChange=1
          killSecond=True
25
26  TerminationSet
27    SpriteCounter stype=goal    win=True
28    SpriteCounter stype=avatar  win=False
```

Listing 6.1: VGDL Definition of the game *Zelda*.

sprites as they appeared in the termination conditions. sword is the only spawned sprite in that game as it only appears when the avatar attacks (avatar spawns a *sword* to attack). The sprite types are assigned based on the interactions of these objects with each other and their sprite class. For example: withkey and nokey are avatar sprites because they are of sprite class ShootAvatar, while monsterQuick, monsterNormal, and monsterSlow are harmful sprites as they can kill the avatar sprites upon collision, as defined in the interaction set.

In the following sections, we describe the sample generators that are provided with the GVGAI framework and competition and other generators that were submitted to the competition in the IJCAI 2016, CIG 2017, and GECCO 2018 or have been published on or before 2018.

6.1.1   SAMPLE GENERATORS

This section discusses the three sample generators provided with the GVGAI framework in detail. Besides discussing how they work and function, we also elaborate at the end about their advantages over each other and show a previous user study that validates our claims.

Sample Random Generator

This is the simplest known generator in the GVGAI framework. The algorithm starts by creating a random sized level directly proportional to the number of game sprite types. Then, it adds a solid border around the level to make sure that all the game sprites will stay inside. Finally, it goes over every tile with 10% chance to add a random selected sprite and makes sure that each game sprite appears at least once and there is only one avatar in the level.

Sample Constructive Generator

This is a simple generator that is provided with the framework that uses the `GameAnalyzer` data to improve the quality of the generated levels. Figure 6.2 summarizes the core steps of the sample constructive generator. The generator consists of four steps, plus pre-processing and post-processing.

1. **(Pre-processing) Calculate cover percentages:** this pre-processing step helps the generator to define the map size and the percentage of tiles that will be covered with sprites. The size of the map depends on the number of game sprites in the game, while the cover percentage is directly proportional with the priority value for each sprite.

2. **Build the level layout:** this is the first step in the generation. The algorithm first builds a solid border around the level then it adds more solid objects inside the level that are connected to each other and not blocking any area.

3. **Add an avatar sprite:** the generator adds a single avatar sprite in any random empty space in the level.

4. **Add harmful sprites:** the generator adds harmful sprites to the map proportional to the distance to the avatar to make sure the player doesn't die as soon as the game starts.

5. **Add collectible and other sprites:** the generator adds collectible and other sprites at random empty locations of the map.

6. **(Post-processing) Fix goal sprites:** the generator makes sure that the number of goal sprites is greater than the specified number of sprites in the termination condition. If this is not the case, the generator adds more goal sprites till this happens.

Sample Genetic Generator

This is a search-based level generator based on using Feasible Infeasible 2-Population genetic algorithm (FI2Pop) [Kimbrough et al., 2008]. FI2Pop is a genetic algorithm that keeps track of two populations: feasible and infeasible populations. The infeasible population contains the chromosomes that do not satisfy the problem constraints. The feasible population on the other hand tries to improve the overall fitness for the problem. At any point during the generation, if

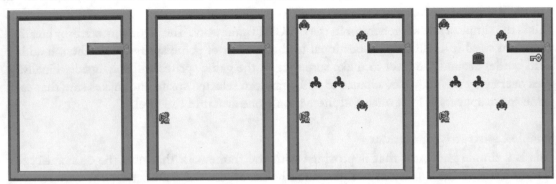

Figure 6.2: Steps applied in the constructive generator for VGDL *Zelda*. Top left: build the level layout. Top right: add an avatar sprite. Bottom left: add harmful sprites. Bottom right: add collectible and other sprites.

the chromosome doesn't satisfy the problem constraints it gets transferred to the infeasible population and vice versa. The initial population is populated using the constructive generator where each chromosome represents a generated level in the form of 2D array of tiles. The generator uses one point crossover around any random tile and three mutation operators.

- **Create:** create a random sprite at an empty tile position.

- **Destroy:** clear all the sprites from a random tile position.

- **Swap:** swap two random tiles in the level.

We use three different game controllers to evaluate the generated levels: a modified version of OLETS, *OneStepLookAhead*, and *DoNothing*. OLETS is the winner of the 2014 single-player planning track of the GVGAI competition [Pérez Liébana et al., 2016]. The algorithm is based on Open Loop Expectimax algorithm (for more details check Chapter 4). The algorithm was modified to play more like a human by introducing action repetition and NIL actions whenever the agent changes direction. This modification was added to influence the generated levels not to require super human reaction time to solve them. *OneStepLookAhead* is just a simple greedy algorithm that picks the best action for the immediate following move. *DoNothing* is a simple algorithm that does not execute any action.

These agents are used as part of the fitness calculation and constraint solving. The feasible population fitness is divided into two parts.

- **Relative Algorithm Performance:** the first part of the fitness calculate the difference in performance between the modified OLETS and *OneStepLookAhead*. This is based on Nielsen et al. [2015], which assumes that a good designed level/game has a high performance difference between a good playing algorithm and bad one playing at it.

- **Unique Interactions:** the second part of the fitness calculates the number of unique interactions that fire in the game due to the player or an object spawned by the player. We hypothesize that a good generated level should have enough interactions that can be fired by the player to keep the game interesting.

For the infeasible population, the algorithm tries to satisfy seven different constraints. These constraints were designed based on our understanding of GVGAI and our intuition about a good level.

- **Avatar Number:** each level has to have only one avatar, no more and no less. This constraint makes sure that there is one controllable avatar when the game starts.

- **Sprite Number:** each level has to have at least one sprite from each non spawned sprites. This constraint makes sure that the level is not missing any essential sprite that might be needed to win.

- **Goal Number:** each level has to have an amount of goal sprites higher than the number specified in the termination condition. This constraint makes sure the level does not automatically end in a victory when the game starts.

- **Cover Percentage:** tiles should only cover between 5% and 30% of each level. This constraint make sure the level is neither empty nor crowded with different sprites.

- **Solution Length:** levels must not be solved in less than 200 steps to make sure the level is not trivial.

- **Win:** levels must be winnable by the modified OLETS to make sure that players can win it too.

- **Death:** the avatar must not die in the 40 steps when using the *DoNothing* player to make sure the player does not die in the first couple of seconds of the game.

Pilot Study and Discussion

In a previous work [Khalifa et al., 2016], the sample generators were compared to each other through a pilot study using 25 human players. The study was conducted on three different VGDL games.

- *Frogs*: is a VGDL port for Frogger (Konami, 1981). The aim of the game is to help the frog to reach the goal without getting hit by a car or drown in water.

- *PacMan*: is a VGDL port for PacMan (Namco, 1980). The aim of the game is to collect all the pellets on the screen without getting caught by the chasing ghosts.

- *Zelda*: is a VGDL port for the dungeon system in The Legend of Zelda (Nintendo, 1986). The aim of the game is to collect a key and reach a door without getting killed by enemies. The player can kill these enemies using their sword for extra points.

For this pilot study, we generated three levels per game by giving each algorithm 5 h maximum for generation. Each user is faced with two generated levels that are selected randomly and they have to decided if the first level is better than the second, the second level better than the first, both are equally good, or both are equally bad.

We hypothesize that the search based levels are better than constructive levels which are better than random generated levels. Table 6.2 shows the result of that study over all the three games. In that table, we only kept the results when any of the two levels is better than the other, removing all the data where both levels are equally good or equally bad. From the table we can be sure that the search based generated levels are better than both constructive and random but it was surprising to find that constructive generator and random generator were on the same tier. From further analysis of the result, we think the result is the constructive generator didn't make sure there is at least one object from every different sprite type which caused some of the constructive levels to be unsolvable compared to the random generated levels.

Table 6.2: Player preferences for each generator aggregated over the three games.

	Preferred	Non-Preferred	Total	Binomial p-value
Search-based vs. Constructive	23	12	35	0.0447
Search-based vs. Random	21	10	31	0.0354
Constructive vs. Random	17	24	41	0.8945

6.1.2   COMPETITION AND OTHER GENERATORS

We can divide the level generator into three main categories based on their core technique: constructive methods, search-based methods, and constraint-based methods. The sample random generator and sample constructive generator are both constructive methods while sample genetic generator is a search-based method. In the following parts, we are going to talk more about these algorithm and their core technique.

Constructive Methods

Constructive methods uses generate levels directly by placing game sprites in the level based on game specific knowledge such as don't place player too close to enemies, don't add solid sprites that isolates areas in the map, etc. These generators don't check for playability after generation as they are supposed to be designed in a way to avoid these problems and make sure the generated content is playable all the time.

Easablade constructive generator: is the winner of IJCAI 2016 level generation competition. This generator uses cellular automata [Johnson, Yannakakis, and Togelius, 2010] on multiple layers to generate the level. The first layer is responsible to design the map layout, followed by the exit sprite and the avatar sprite, then the goal sprites, harmful sprites, and others.

N-Gram constructive generator: was submitted to CIG 2017 level generation competition. It uses a n-gram model to generate the level. The generator uses a recorded playthrough to generate the levels. The algorithm has cases for each different type of interactions, for example: attacking involves adding an enemy some where in the level, walking around a certain tile involve adding solid objects, etc. The n-gram model is used to specify these rules so instead of reacting to a single action, the system responds to a n-sequence of actions. During the generation the algorithm also keeps track of all the placed sprites to make sure it doesn't overpopulate the level. Then, the avatar sprite at the bottom center of the level is added.

Beaupre's constructive pattern generator: is a constructive algorithm that was designed for Beaupre et al. [2018] work on analyzing the effect of using design patterns in automatic generation of levels for the GVGAI framework. In their work, they analyzed 97 different games from the GVGAI framework using a 3×3 sliding window over all the levels after transforming the levels to use sprite type from the *GameAnalyzer* instead of the actual sprites. They constructed a dictionary that contains all these different patterns (they discovered 12,941 unique patterns) and classified it based on the type of different objects mixed in. There are border patterns which are patterns that appear on the border of the level, avatar patterns which are patterns that contain an avatar sprite in them, and others. The algorithm starts by checking if the game have solid sprites. If that was the case, it fills the border areas using border patterns. The rest of the level is picked randomly from the rest of the patterns based on their distribution while making sure there is only avatar in the level and the level is still fully connected. Finally, the system converts the generic sprite types to specific game sprites while making sure that goal sprites numbers are larger than the specified number in the termination condition.

FrankfurtStudents constructive generator: is designed for GECCO 2018 level generation competition. This generator is more handcrafted generator that tries to identify the type of the game (racing game, catching game, etc.) and based on that type defines the level size and gives each game sprite a preferred position with respect to other sprites and the level.

Search-Based Methods

Search-based methods uses a search based algorithm such as genetic algorithm to find levels that are playable and more enjoyable than random placement of sprites. This section describes all the known search-based generators.

Amy12 genetic generator: is build on top of the sample genetic generator. This generator was submitted to IJCAI 2016 competition. The idea is to generate level that has a certain suspense curve. Suspense is calculated by measuring the number of actions that leads to death during the gameplay using OLETS agent. The algorithm tries to modify the level to get a suspense curve with three peak points of less than 50% height. The advantage of doing that that it makes sure that the generated levels are winnable as levels that are not winnable will have higher peaks in the suspense curve.

Jnicho genetic generator: uses a standard genetic algorithm compared to FI2Pop used in the sample generator [Nichols, 2016]. The generator combines the constraints and relative algorithm performance in a single fitness function where the relative algorithm performance is measured between a MCTS agent and an OneStepLookAhead agent. The score values of these agents is normalized between 0 and 1 to make sure that the relative algorithm performance doesn't overshadow the rest of the constraints. This generator was submitted to IJCAI 2016 competition.

Number13 genetic generator: is a modified version of the sample genetic generator. This generator was submitted to IJCAI 2016 competition. The generator uses adaptive methods for crossover and mutation rates with a better performing agent than the modified OLETS. It also allows crossover between feasible and infeasible chromosomes which was not allowed in the sample generator.

Beaupre's evolutionary pattern generator: uses similar idea to the their constructive generator described in the constructive methods. This generator is a modified version of the sample genetic generator provided with the framework but with the chromosome represented as 2D array of patterns instead of tiles. In that case, they use their constructive approach to initialize the initial population.

Sharif's pattern generator: is submitted as a participant in GECCO 2018 competition. This generator similar to Beaupre's evolutionary pattern generator in using identified patterns to build the level. In a previous work by Sharif, Zafar, and Muhammad [2017], they identified a group of 23 different unique patterns that they are using during the generation process. These patterns are selected to have similar meaning to the pattern identified in the work by Dahlskog and Togelius in generating levels for Super Mario Bros (Nintendo, 1985) [Dahlskog and Togelius, 2012].

Architect genetic generator: is the winner of GECCO 2018 level generation competition. The algorithm is build on the sample genetic generator provided with the framework, the only difference that it uses two-point crossover and a new constructive initialization technique. The new constructive technique is similar to the sample random generator with some improvements. It starts with calculating the size of the map, then building a level layout similar to the one used in the sample constructive technique, followed by adding an avatar to the map. Finally, 10% of the map is picked randomly from all the possible sprites.

Luukgvgai genetic generator: is another GECCO 2018 submission, similar to the other is a modified version of the sample genetic generator using tournament selection and two-point crossover.

Tc_ru genetic generator: is a modified version of the sample genetic generator that uses eigth-point crossover and tournament selection instead of rank selection. It also used cascaded fitness where the relative algorithm performance has higher precedence than the unique interactions. This algorithm was submitted to GECCO 2018 GVGAI level generation competition.

Constraint-Based Methods

Constraint-based methods used a constraint solver to generate levels. The idea of the generator is to find the right constraints needed to make sure the generated content is the targeted experience. So far only one generator has been developed for GVGAI using that technique.

ASP generator: Neufeld, Mostaghim, and Pérez Liébana [2015] uses answer set programming (ASP) to generate levels. The ASP programs are evolved using evolutionary strategy that uses relative algorithm performance between the *sampleMCTS* agent and the *sampleRandom* agent. The evolved rules are divided into three different types. The first type is a basic set of rules that make sure the generated level is not complicated such as making sure there is only one sprite per tile. The second type are more game specific rules based such as identifying the Singleton sprites in the current game. The third type is concerned with the maximum number of objects that can be produced for a certain sprite type.

6.1.3 DISCUSSION

The presented generators differ in the amount of time needed to generate a level and the features of the generated content. The constructive generators take the least amount of time to generate a single level without a guarantee that the generated level is beatable. On the other hand, both search-based and constraint-based generators take longer time but generate challenging beatable levels as they use automated playing agents as a fitness function. The constraint-based generator only takes long time to find an ASP generator which could be used to generate many different levels as fast as the constructive generators, while search-based generators take a long time to find a group of similar looking levels.

Some of the presented algorithms were submitted to the GVGAI level generation track in IJCAI 2016 (Easablade, Amy12, Number13, and Jnicho genertors), CIG 2017 (N-Gram constructive generator), and GECCO 2018 (Architect, FrankfurtStudents, Sharif, Luukgvgai, and Tc_ru generators). These generators were compared with respect to each other during the competition.

Figure 6.3 shows the results from IJCAI 2016 competition.

The algorithms were tested against four GVGAI games:

- **Butterflies:** is a VGDL game about collecting all butterflies before they open all the cocoons in the level.

- **Freeway:** is a VGDL port of Freeway (David Crane, 1981). Similar to Frogs, the goal is to reach the other side of the road without being ran over by cars. The difference is that this game is a score-based game where the player need to reach the goal more than one time to get a better score.

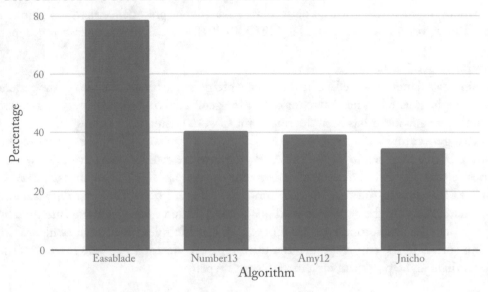

Figure 6.3: IJCAI 2016 level generation track results.

- **Run:** is a VGDL runner game where the player is trying to outrun a flood coming from behind and reach the exit door in time.

- **The Snowman:** in this game, the goal is to stack the snowman pieces (legs, trunk and head) in the correct order to create a snowman.

The results are very clear that Easablade beats the other three generators. A possible reason is that Easablade generated a few amount of sprites in the generated scene with a nice layout. On the other hand, all the generated levels were either unplayable (exits hidden behind a wall) or easy to beat (exits directly beside the player).

Figure 6.4 shows the results from GECCO 2018 competition. The algorithms were tested against three GVGAI games picked from Stephenson et al. [2018] work on finding the most discriminatory VGDL games.

- **Chopper:** is a VGDL action game where the player needs to collect bullets to kill the incoming tanks without getting killed by the bullets from the enemy tanks. Also, the player has to protect the satellites from being destroyed.

- **Labyrinthdual:** is a VGDL puzzle game where the player tries to reach the exit without getting killed by touching spikes by changing their color to pass through colored obstacles.

- **Watergame:** is a VGDL puzzle game about trying to reach the exit without drowning in water. To pass the level, the player has to push couple of potions around to convert the water to floor so the player can pass over it.

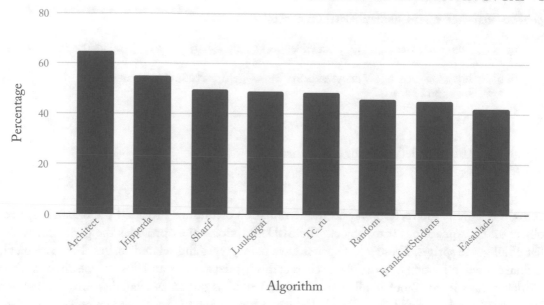

Figure 6.4: GECCO 2018 level generation track results.

Architect won the competition with a small difference compared to Easablade. The reason behind the winning is not clear as some other generators have similar looking levels. Shockingly, Easablade got the worst rating compared to the random generator. We think that the choice of games affected the outcome of the competition. For example a puzzle game like Watergame is only playable if it is solvable having a big level and sparse object will be against the generator interest while in a game like freeway that might be better.

6.2 RULE GENERATION IN GVGAI

The rule generation track was introduced in 2017 as a new competition track [Khalifa et al., 2017]. It is the second-generation track for the GVGAI framework. The competition focuses on creating a new playable games for a provided game level with all the game sprites. In this track, every participant submits a rule generator that produces the interaction sets and termination conditions in a fixed amount of time. The framework provides the generator with a game level in form of 2D matrix of characters which can be translated to its corresponding game sprites using the provided level mapping. In return, the generator produces two arrays of strings that cover the game interaction set and termination conditions.

The game level and the game sprites are provided through the `SLDescription` object. Figure 6.5 shows the `AbstractRuleGenerator` class the user needs to extend to build their own rule generator. The user has to implement `generateRules` function and return the two arrays that

```
public abstract class AbstractRuleGenerator {

    public abstract String[][] generateRules(SLDescription sl, ElapsedCpuTimer time);

    public HashMap<String, ArrayList<String>> getSpriteSetStructure(){
        return null;
    }
}
```

Figure 6.5: AbstractruleGenerator class functions.

contain the game interaction set and the termination conditions provided the SLDescription object and ElapsedCpuTimer object. The SLDescription object provides the generator with a list of all game sprites. These game sprites have name, type, and related sprites. For example: in Zelda, shown in Listing 6.1, the player character is named avatar and it is of type ShootAvatar (which means it can shoot in all four directions) and has sword as related sprites (as it shoots it in any of the four directions). Beside the game sprites, the SLDescription object provides a 2D matrix that represents the game level where each cell represents the sprites that are present in that cell. When running the competition, the ElapsedCpuTimer provides the generator with five hours to finish generation.

The user can also override an optional function called getSpriteSetStructure which returns a hashmap between a string and an array of strings. The hashmap represents the sprite hierarchy required during the rule generation. For example: in zelda, the user can generate three rules that kill the *avatar* when it hits any monsterQuick, monsterNormal, and monsterSlow sprites or they can generate one rule that kills the avatar when it hits a harmful sprite and then define the harmful sprite in the sprite structure as all these three monsters.

The framework also provides the user with a LevelAnalyzer object that analyzes the provided level and allows the user to ask about sprites that cover certain percentages of the map and/or of a certain type. For example: the generator can get the background sprite by asking the LevelAnalyzer to get an Immovable sprite that covers 100% of the level. This information could be used to classify the game sprites to different classes. Table 6.3 shows some example classes that can be recognized from the game of Zelda presented in Figure 6.6. These categories are defined based on general game knowledge. For example: score objects such as coins in Mario covers small percentage of the level.

We ran the rule generation track twice: at CIG 2017 and GECCO 2018. In both times, we didn't get any submissions due to the lack of advertisement for the track and the harder the problem is to tackle (generating rules for a game is harder than just producing a level). In the following sections, we will discuss the sample generator that were provided with the competition and the only other generator that we found in the literature which was created before the track existed.

Table 6.3: The *LevelAnalyzer* data for the game of *Zelda*

Class Type	Sprite Image	Sprite Type	Threshold	Surrounding Level
Background		Immovable	≥ 100%	FALSE
Wall		Immovable	< 50%	TRUE
Score/Spike		Immovable	< 10%	FALSE

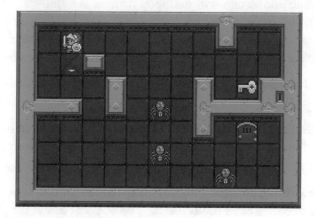

Figure 6.6: The provided level for Zelda.

6.2.1 SAMPLE GENERATORS

In similar manner to the level generation track, the rule generation track comes with three different generators: random, constructive, and search based generators. These generators increase in complexity and quality of the generated content.

Sample Random Generator

The sample random gene rator is the simplest of all the generators. The generator just pick any random interactions and termination conditions that will compile into a VGDL game without any errors. The generator follows the following steps to generate a random VGDL game.

1. Pick a random number of interactions.

2. Generate a random amount of interactions by repeating the following steps until reaching the chosen amount of interactions while making sure it compiles with no errors:

    (a) pick two random sprites.

    (b) pick a random `scoreChange` value.

    (c) pick a random interaction.

3. Generate two terminal conditions for winning and losing.

    • *Winning:* either a time out with a random time or a certain sprite count reaches zero.

    • *Losing:* if the avatar dies, the game is lost.

Sample Constructive Generator

The constructive generator is more sophisticated than the sample generator. It uses a template based generation to generate a good playable game. The generator has a game template that was designed based on game knowledge. For example: if there is a non-playable character (NPC) running after a certain object, there is a high chance that it will kill it upon collision. The generator utilizes the sprite types from `SLDescription` object and sprite classes from `LevelAnalyzer` object to fill the template. The following are the steps taken by the sample generator.

1. **Get Resource Interactions:** collectResource interaction is added for all the resource sprites.

2. **Get Score and Spike Interactions:** each object has 50% chance to be either collectible or harmful. Collectible sprites are killed upon collision with the avatar and give one point score. Spike sprites kills the avatar upon collision.

3. **Get NPC Interactions:** different interactions are added for different types of NPCs. Usually they are either collected by the avatar for 1 score point or kill the player upon collision. Some NPCs could spawn other sprites which also could be either collectible or deadly.

4. **Get Spawner Interactions:** similar to NPCs, spawner sprites decide if the generated sprites will kill the avatar or get collected for points.

5. **Get Portal Interactions:** if the portal is of type *door*, the avatar will destroy it upon collision. Otherwise, the avatar is moved to the destination portal.

6. **Get Movable Interactions:** the generator decides randomly if these movable objects are harmful or collectible sprites.

7. **Get Wall Interactions:** the generator decides if the wall sprites will be fire walls or normal walls. Fire walls kill any movable sprite upon collision, while normal walls prevent any movable sprite.

8. **Get Avatar Interactions:** this step only happens if there are any harmful sprites being generated and the avatar can shoot bullets. The generator adds interactions between the bullets and harmful sprites to kill both of them upon collision.

9. **Get Termination Conditions:**

- *Winning:* the generator picks a random winning condition such as the avatar reached a door, all harmful sprites are dead, all the collectible sprites are collected, or time runs out.

- *Losing:* the game is lost when the avatar dies.

Sample Genetic Generator

Similar to the level generation track, the search based algorithm uses the FI2Pop algorithm [Kimbrough et al., 2008] to generate interaction set and termination conditions. The infeasible chromosomes try to become feasible by satisfying these three constraints.

- Interaction set and termination conditions compile with no errors.

- A do-nothing agent stays alive for 40 frames.

- The percentage of bad frames are less than 30%, where bad frames are frames where one sprite or more are outside of the screen boundaries.

If the chromosome satisfies these constraints, it get moved to the feasible population. The feasible chromosomes try to improve their fitness. The feasible fitness consists also of three parts.

- Increase the relative algorithm performance [Nielsen et al., 2015] between three different agents (OLETS agent, MCTS agent, and random agent, in this order of performance).

- Increase the number of unique rules that are fired during the game playing session. Game rules are added to be used, if a rule is not being used by any agent then it violates this constraint.

- Increase the time the agent takes to win the level. We don't want games that could be won in less than 500 frames (20 s).

We used rank selection with 90% chance for crossover and 10% chance for mutation. We used one-point crossover to switch both interaction set and termination conditions. For the mutation, three different operators were used.

- **Insertion:** either inserting a new rule/termination condition or inserting a new parameter.

- **Deletion:** either delete a rule parameter or delete an entire rule/termination condition.

- **Modify:** either modify one of the rule parameters or modify the rule/termination condition itself.

We initialized the algorithm with 10 constructive chromosomes, 20 random chromosomes, and 20 mutated versions of the constructive chromosomes to a have a population of size of 50. We used 2% elitism which keeps the best chromosome between generations.

Pilot Study and Discussion

We applied these three generators on three different games (Aliens, Boulderdash, and Solarfox) selected based on the different algorithm performance by Bontrager et al. [2016] work. Our first study was to see the diversity of the generated content between these generator on these three games. Figure 6.7 shows the probability density function of how similar the generated games to each other where 0.0 means 100% similar and 1.0 means they are totally different. These distribution were calculated by generating 1,000 games from both sample random generator and sample constructive agent and 350 generated games from the sample genetic generator (due to time constraints). From all distributions, we can see that the sample genetic generator is able to generate games that ranges from very similar to each other to totally different, while the constructive generator was very limited to similar games, which was expected as we are using a template with a small amount of parameters to be changed.

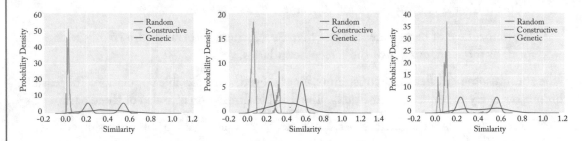

Figure 6.7: Probability density of the similarity metric between the generated games. These graphs represent Aliens, Boulderdash, and Solarfox from left to right. "0" means the games are identical, while "1" means the games are totally different.

We also conduced a similar user study to the level generator track to compare these three generators to each other in terms of preference. The users were faced with two generated games with the same level layout from either the same generator or a different generator, and we asked them to pick which one they thought it was best (first or second), both of them being equally good, or neither of them being good. Table 6.4 shows the results of the user study. As we expected constructive generator and genetic generator are mostly preferred over the random generator except for Aliens in the genetic generator. The reason for aliens being different is the genetic generator didn't have much time to evolve so to satisfy the constraints it generated a game with undoAll as an interaction between player and background to satisfy the bad frames constraint. undoAll interaction pauses the whole game, not allowing for anything to happen. Another remark is that the genetic generator was not preferred over the constructive one. We think the reason for that is the constructive templates were better designed and users never noticed they are similar to each other.

Table 6.4: Comparison between our rule generators where "Genetic" is the sample genetic generator, "Rnd" is the sample random generator, or "Const" is the sample constructive generator. The first value is the number of times the user preferred the first game over the second. The second value is the total number of comparisons.

	Aliens	Boulderdash	Solarfox
Genetic vs. Random	2/8	7/7	11/15
Genetic vs. Constructive	0/14	8/14	6/18
Constructive vs. Random	9/10	10/11	4/5

6.2.2   OTHER GENERATORS

As far as we are aware, there is only one other work that was done toward rule generation, by Nielsen et al. [2015]. The generator is similar to the sample genetic generator as the sample generator idea was based on that work. Both generators use the difference between the performance of different algorithms (relative algorithm performance) as an evaluation function. The difference is that this generator uses evolutionary strategies with mutation operators to generate an entire game instead of interaction set and termination conditions.

6.2.3   DISCUSSION

The rule generation competition has been running for two years but no one has submitted yet to it. We think the reason behind that might be that the competition is harder than level generation (need more computations) and usually runs at the same time with level generation so the competitors either choose to participate in the level generation competition. We think that there is a huge opportunity of research in rule generation competition. Some researchers work on generating full games [Togelius and Schmidhuber, 2008] based on Ralph Koster Theory of Fun [Koster, 2013], generating games by modifying the game code [Cook et al., 2013], and generating games based on a single work [Cook and Colton, 2014], etc.

6.3   EXERCISES

The GVGAI Framework is available in a Github repository.[1] Use the release 2.3[2] in order to run the same version presented here. This chapter proposes two procedural content generation challenges hosted by the GVGAI competition.[3]

Projects can be proposed within this context easily:

---

[1]https://github.com/GAIGResearch/GVGAI
[2]https://github.com/GAIGResearch/GVGAI/releases/tag/2.3
[3]These exercises are also available at this book's website: https://gaigresearch.github.io/gvgaibook/.

- Build a level generator that improves the sample level generators provided by the framework. You can run the level generators from the class
`tracks.levelGeneration.TestLevelGeneration`.

- Analogously, you can build a rule generator from the sample ones provided. You can run the rule generation test from the class `tracks.ruleGeneration.TestRuleGeneration`.

- For both cases above, inspiration can be found in Procedural Content Generation in Games, by Shaker, Togelius, and Nelson [2016] You can also submit your generator(s) to the yearly editions of the GVGAI competition.

- What is beyond rule generation? Could you create a generator that creates *complete* games in VGDL?

CHAPTER 7

Automatic General Game Tuning

Diego Pérez Liébana

7.1 INTRODUCTION

Automatic Game Tuning refers to an autonomous process by which parameters, entities or other characteristics of games are adjusted to achieve a determined goal. This goal could be to favor a specific type of player, create games that are more balanced in the case of two-or-more player games, or in order to provide a determined game playing experience to the players. The fact that is *automatic* allows to save some valuable time to game designers and testers, who otherwise need to manually change and play-test the different versions of the game. Furthermore, it can serve as a way of finding new variants of a game that no human had thought before.

As mentioned previously, the fact that the VGDL is the backbone of GVGAI, allows for a fast tweak-and-test process for game tuning. This chapter studies how VGDL has been modified to facilitate automatic game tuning, by including variables in the language that can be modified from the engine. With this, the GVGAI framework allows not only to create content (levels or rules) for games, as seen in the previous chapter, but also to work on VGDL game *spaces*. This parameterization of GVGAI games (see Section 7.2 in this chapter) exposes a game space for algorithms to search in; this is, the collection of all possible game variants that the values of the different parameters facilitate.

One possibility to navigate these game spaces is to use optimization techniques. Given an objective (or fitness function), the optimization method explores values for the exposed parameters progressively in order to find games that fulfil better the desired criteria. In this chapter, we propose, in Section 7.3, to use the N-tuple bandit evolutionary algorithm (NTBEA) for this purpose. We then focus on automatic game tuning with two different objectives. First, to show how a game can be tweaked to favor one player over another one (see Section 7.3.2). Second, in Section 7.4, we modify several games to offer different experiences to the player—concretely in the way and time score opportunities are presented along the game.

7.1.1   PREVIOUS WORK

One of the first examples of research on game spaces that can be found in the literature [Isaksen et al., 2015] defined a parameter space for the game *Flappy Bird*. In this game, the player has two actions to execute at every frame (*tap* or *no tap*). Each tap makes the player (a bird) to flap its wings and gain some height, while no tapping makes it fall due to the force of gravity. The player must travel to the right of the screen, going through a series of gaps between pipes without touching them. The game ends when the bird touches one of these pipes. Evolutionary algorithms were used to explore the values of the game parameters and the different games that resulted of changing them. Examples of these parameters are the pipe lengths and widths, the (horizontal) distance between pipes, player size, force of gravity, etc. The authors identified four different settings, each one providing a unique gameplay experience. A follow-up work aimed at finding different difficulty levels [Isaksen, Gopstein, and Nealen, 2015] showed that all playable games are clustered in a certain part of the search space.

In a similar way, Liu et al. [2017] proposed game parameterezation for the two-player game *Space Battle*, in which two ships move in real-time in a 2D and wrapped space while shooting at each other. The authors used a random mutation hill climber (RMHC) to evolve game parameters such as the maximum ship and missile speeds, cooldown time (between shootings), missile cost, ship radius and thrust power. The objective of this work is to find game variants in which an MCTS player would defeat rotate and shoot (RAS), which is a very simple but powerful strategy in this game. Using the UCB1 (3.1) to select which parameter RMHC should mutate next resulted to be an efficient way to explore the search space and find games with an interesting skill-depth.

A later work by Kunanusont et al. [2017] used a novel Evolutionary Algorithm, the *N-Tuple Bandit EA* (or NTBEA [Lucas, Liu, and Pérez Liébana, 2018c]) to explore an even larger space in a new version of *Space Battle*, which counted on 30 in-game parameters. The aim was to find games that would favor skilled players against weak ones, using RAS and two GVGAI agents for training: MCTS (strongest player) and One-Step Lookahead (1SLA, the weakest). The fitness of each game was calculated as the minimum gap of performance between the pairs MCTS-RAS and RAS-1SLA.

7.2   GVGAI PARAMETERIZATION

The first step required to explore game spaces in VGDL games is to adapt the language to account for game parameterization. In order to do this, we enhanced VGDL in two ways: first, defining a new section for the language (*ParameterSet*) that would define the types and values of all the parameters that can exist in a VGDL game. Then, allowing the possibility of defining *variables* as values for the properties listed in the *SpriteSet*, *InteractionSet* and *TerminationSet*.

Listing 7.1 shows an example of a VGDL game, Alens, that has been parameterized. In this case, only the *SpriteSet* has been modified (see the default VGDL code for this game at List-

```
 1  GameSpace square_size=32
 2   SpriteSet
 3    background > Immovable img=oryx/space1 hidden=True
 4    base      > Immovable img=oryx/planet
 5    avatar    > FlakAvatar  stype=sam img=oryx/spaceship1
 6    missile > Missile
 7     sam    > orientation=UP speed=SSPEED singleton=IS_SAM_SINGLE img=
              oryx/bullet1
 8     bomb > orientation=DOWN speed=BSPEED img=oryx/bullet2
 9    alien    > Bomber stype=bomb prob=APROB cooldown=ACOOL speed=ASPEED

10     alienGreen > img=oryx/alien3
11     alienBlue > img=oryx/alien1
12    portal   > invisible=True hidden=True
13     portalSlow > SpawnPoint stype=alienBlue cooldown=PCOOL total=
              PTOTAL
14
15   ParameterSet
16    #{Name} > {values(min:inc:max)/(bool)} {desc_string} {[opt]val}
17
18    SSPEED    > values=0.1:0.1:1.0      string=Sam_Speed        value=0.5

19    BSPEED    > values=0.1:0.1:1.0      string=Bomb_Speed
20    APROB     > values=0.01:0.05:0.75   string=Alien_Bomb_Probability
21
22    ACOOL     > values=1:1:5            string=Alien_Cooldown
23    ASPEED    > values=0.5:0.1:1.0      string=Alien_Speed
24    PCOOL     > values=1:1:5            string=Alien_Portal_Cooldown
25    PTOTAL    > values=10:5:60          string=Alien_Portal_Total
26
27    IS_SAM_SINGLE > values=True:False string=Is_Sam_Singleton
```

Listing 7.1: VGDL Definition of the game *Aliens* enhanced with a ParameterSet. InteractionSet and LevelMapping are unchanged with respect to the default VGDL code for this game (see Listing 2.1).

ing 2.1) and a *ParameterSet* has been added to define the new parameters. Finally, it is important to notice that the first keyword of the VGDL game description file must be GameSpace.[1]

Each line of the *ParameterSet* (lines 15–27) define a new parameter that can be used in the other sets of the description. Each one of these parameters can have up to four different fields.

- **Name:** this is the variable name, a single word that must be used in the other sets to reference this parameter.

- **Values:** this field defines the possible values this parameter may take. Three types are allowed and they are implicitly defined when providing the values, two numeric (int and double) and a boolean.

---

[1]Note that, in a fully defined VGDL game, this is *BasicGame*.

– **Numeric**: values are indicated in a tuple of three numbers, $a : b : c$, where a is the minimum, b the increment and c the maximum. For instance, line 19 defines the parameter *BSPEED*, which can take values as defined in 0.1 : 0.1 : 1.0, where the minimum value is 0.1, the maximum 1.0 and the increment is 0.1. This determines that all possible values for this parameter are 10: 0.1, 0.2, 0.3, 0.4, 0.5, 0.6, 0.7, 0.8, 0.9, and 1.0.

– **Boolean**: these parameters are always defined as *True* : *False* and can take only these two values. An example of a boolean parameter can be seen in line 27.

• **Descriptive String**: an easy to read expression that describes (better than the variable name) what the parameter does.

• **Value**: this is an optional field that allows the designed to initialize a value for the parameter before running the game (see line 18 for an example). If this is present, the values field is ignored. The objective of this field is to facilitate play-testing with the parameters.

Once the *ParameterSet* is defined, it is possible to add variables to the other sets. Listing 7.1 shows a few examples of this. For instance, line 9, which defines the properties of the aliens (enemies) of this game, uses three variables defined in the *ParameterSet*: *APROB* (which determines the probability of an alien dropping a bomb at each frame), *ACOOL* (which decides the number of consecutive frames while the aliens do not execute any action) and *ASPEED* (speed of the aliens, measured in grid cells per movement action).

The framework permits creating games from game space definitions by providing an array of integers for the values of the parameters. Listing 7.2 shows an example in GVGAI to initialize game spaces. Each one of the values in the array (i) maps one parameter, so the value of this parameter will be $lower_{bound} + i * increment$. The order of the elements of this array is the same as the order listed when using the function `printDimensions()`, from the object `DesignMachine`.

For the game space of Aliens defined above, the call to `printDimensions()` provides the following output, which also include the size of the space of possible instantiations of this game and of each dimension:

This feature of GVGAI is available for both single- and two-player games. In the rest of this chapter, we show two different use cases in which evolution is used to find different instantiations according to certain search criteria.

7.3   EVOLVING GAMES FOR DIFFERENT AGENTS

In this section, we propose the following problem: given two agents with different skill capabilities, is it possible to find instances of one game (Aliens) in which one agent performs better than the other? And vice versa?

First, we analyze a new Evolutionary Algorithm, the N-Tuple Banding Evolutionary Algorithm [Lucas, Liu, and Pérez Liébana, 2018b], which will be used to search the space of

```
1  //Reads VGDL and loads game with parameters.
2  String game = "game_filename.txt";
3  DesignMachine dm = new DesignMachine(game);
4
5  //1) Creating a new instantiation of the game space at random:
6  int[] individual = new int[dm.getNumDimensions()];
7  for (int i = 0; i < individual.length; ++i)
8      individual[i] = new Random().nextInt(dm.getDimSize(i));
9
10 //We can print a report with the parameters and values:
11 dm.printValues(individual);
12
13 //Play the game (a human in control)
14 dm.playGame(individual, game, level1, seed);
15
16
17 //2) Creating a new individual with specific values:
18 //   Each parameter will take a value = "lower_bound + i*increment"
19 individual = new int[]{2, 2, 0, 4, 8, 3, 9, 4};
20
21 dm.playGame(individual, game, level1, seed);
```

Listing 7.2: GVGAI Code example (in Java) game space initialization.

```
1  Individual length: 8
2  Value    S(D)   Range           Description
3  0.3      10     0.1:0.1:1.0     Bomb_Speed
4  0.7      6      0.5:0.1:1.0     Alien_Speed
5  false    2      True:False      Is_Sam_Singleton
6  5        5      1:1:5           Alien_Cooldown
7  50       11     10:5:60         Alien_Portal_Total
8  0.4      10     0.1:0.1:1.0     Sam_Speed
9  0.46     15     0.01:0.05:0.75  Alien_Bomb_Probability
10 5        5      1:1:5           Alien_Portal_Cooldown
11 Search Space Size: 4.950E6
```

Listing 7.3: Game Space for the game aliens, as indicated in the GVGAI output.
Columns are, from left to right: Final value, Dimension size, Range of values and
Comprehensive description.

possible Alien games. Second, we will observe the different results that can be achieved with
this method in this environment.

7.3.1   THE N-TUPLE BANDIT EVOLUTIONARY ALGORITHM

The N-tuple bandit evolutionary algorithm (NTBEA) [Lucas, Liu, and Pérez Liébana, 2018b]
is an optimization algorithm specially suited for large search spaces in which the evaluation of
each one of its points is computationally very expensive. NTBEA counts on an N-Tuple system
that captures the statistics of the inherent model and the combination of the values of its discrete
parameters.

NTBEA is formed of three different parts: a bandit landscape model, an evolutionary algorithm and a fitness evaluator subject to noise. In principle, we can assume that the execution of querying the landscape model is negligible in comparison with evaluating a potential solution. Figure 7.1 depicts the three components of the algorithm.

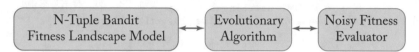

Figure 7.1: Key components of NTBEA.

NTBEA works as follows. Search starts from a single point in the search space, chosen uniformly at random. We refer to this point as the *current point*. This point is evaluated once in the target problem, using the noisy evaluator. There is no need for resampling (even in noisy problems) directly. The bandits built in the model may require to re-evaluate a candidate solution in future steps. Note that the algorithm also works for noise-free problems, but without loss of generality it assumes the harder case (where the problem is noisy).

The current point and its fitness value is stored in the bandit landscape model (for brevity, referred to as *the model* from now on). Then, the algorithm advances to select the next current point. In order to do this, the model is then searched around the neighborhood of the current point. This neighborhood is defined by a number of neighbors and the proximity distribution to the current point. This distribution is controlled directly by the mutation operator. The next point will be such point in the neighborhood with the highest estimated Upper Confidence Bounds value (see UCB1, Equation (3.1)). This process continues until a the termination condition is met (i.e., evaluation budget or some other criterion).

Estimating UCB Values: an N-Tuple Approach

One of the key parts of this algorithm is how to estimate the UCB values when sampling a large search space. With large, we assume that it is impossible to evaluate all possible points in the solution space, because the number of fitness evaluation allowed by the budget is smaller than the search space size. The relationship between the points sampled and their neighbors in the search space needs then to be modeled properly:

$$a^* = \operatorname*{argmax}_{a \in A(s)} \left\{ Q(s, a) + C \sqrt{\frac{\ln N(s)}{N(s, a) + \epsilon}} \right\}. \tag{7.1}$$

UCB1, shown again in Equation (7.1), combines a exploitation term ($Q(s, a)$ on the left) and an exploration expression (on the right) balanced by a control parameter C (higher values lead to a more exploratory search and lower values produce more greedy selections). $N(s)$ is the total number of times a bandit has been played, while $N(s, a)$ indicates the number of times the arm a has been played.

For NTBEA, each dimension of the search space is modeled as an independent multi-armed bandit and each arm represents a possible value. The standard UCB formula does not contain ϵ, in order to guarantee that all arms are pulled at least once. In our case, this would be impractical, as it would force an exhaustive exploration of the search space. ϵ in Equation (7.1) relaxes this requirement.

Additionally, combinations of arms are also modeled as *super-bandits*. In a d-dimensional search space where each dimension has n possible values, the largest super-bandit (which combines all dimensions) would have n^d arms. Rather than using such large bandit, we aggregate over all the N-Tuples in the N-Tuple System model. Let N be the N-Tuple indexing function such that $N_j(x)$ indexes the jth bandit for a point x in the search space. The aggregate UCB value for solution point x is computed as the unweighted average, as defined in Equation (7.2), where m indicates the total number of bandits defined:

$$v_{UCB}(x) = \frac{1}{m} \sum_{j=1}^{m} \text{UCB}_{N_j(x)}. \qquad (7.2)$$

The N-Tuple bandit system sub-samples the dimensions of a d-dimensional search space with a number of N-tuples. N can takes any values from 1 to d, reaching 2^d bandits if all tuples are considered. Each N-Tuple is assigned a look-up table (LUT) that stores statistical summaries of the values associated with it. These statistics contain the number of samples, its sum and the sum of the square of the fitness of these samples. These values allow for the computation of the mean, standard deviation and standard error fore each N-Tuple. These measurements not only provide a great insight into the system that is being modeled, but they also provide all components required for Equations (7.1) and (7.2).

NTBEA is described in Algorithm 7.7. First, it chooses a random point in the serach space (*current point*). Each one of these points is represented with a vector of integers, where each element is an index to a value in that dimension. There is no restriction toward the type of value it refers to (i.e., integer, double, boolean, etc.).

Given an evaluation budget, the following steps are repeated until the search finishes.

1. A (noisy) fitness evaluation of the current point is made, to then store its fitness in the N-Tuple Fitness Landscape Model. This is the value given to that solution point (lines 6 and 7 in Algorithm 7.7).

2. From the current solution, a set of unique neighbors is generated using the mutation operator. This represents the population of the current iteration (line 8).

3. The fitness landscape model calculates the (ucb) value for each one of these neighbors, using Equations (7.1) and (7.2). The new current solution is set as the neighbor with the highest UCB value.

Algorithm 7.7 The N-Tuple Bandit Evolutionary Algorithm. This description outlines the simplest case (1 current point), but a population-based version is also possible.

Input: $n \in N^+$: number of neighbors
Input: $p \in (0, 1)$: mutation probability
Input: $flipOnce \in \{true, false\}$ indicates if flip at least once or not during mutation
Output: $LModel$: landscape model of the problem.

1: **procedure** NTBEA $(n, p, flipOnce)$
2:   $t = 0$ {Counter for fitness evaluations}
3:   $LModel \leftarrow \varnothing$ {Initialize the fitness landscape model}
4:   $current \leftarrow$ random point $\in \mathbb{S}$
5:   **while** $t < nbEvals$ **do**
6:     $value \leftarrow fitnesscurrent$
7:     add $< current, value >$ to $LModel$
8:     $Population \leftarrow$ Neighbors$LModel, current, n, p, flipOnce$
9:     $current \leftarrow \arg\max_{x \in Population} v_{UCB}(x)$
10:    $t \leftarrow t + 1$
11: **end while**
12: **return** $LModel$
13:
14: **procedure** Neighbors $(model, x, n, p, flipOnce)$
15:   $Population \leftarrow \varnothing$ {Initialize empty set}
16:   $d \leftarrow |x|$ {Get the dimension}
17:   **for** $k \in \{1, \ldots, n\}$ **do**
18:     $neighbor \leftarrow x$
19:     $i \leftarrow 0$
20:     **if** $flipOnce$ **then**
21:       $i \leftarrow$ randomly selected from $\{1, 2, \ldots, d\}$
22:       **for** $j \in \{1, \ldots, d\}$ **do**
23:         **if** $i == j$ **or** $RAN < p$ **then**
24:           Randomly mutate value of $neighbor_j$
25:         **end if**
26:       **end for**
27:     **end if**
28:     Add $neighbor$ to $Population$
29:   **end for**
30:   **return** $(Population)$

Once the budget has been exhausted, NTBEA recommends a point among the evaluated solutions and their neighbors, in which each dimension is set to the value with maximal approximate value defined in (7.2).

An Illustrative Example

We take a 5-dimensional space and model it using five 1-tuples and one 5-tuple. In this model, we have four sample points (three of them unique) and their fitness as shown in Figure 7.2 (left). Given these fitness values, the first 1-tuple (i.e., corresponding to the first dimension) has a LUT entry with two entries with a mean of 1 for $LUT[0****]$ and $\frac{2}{3}$ for $LUT[1***]$. The 5-tuple has three non-empty entries: $LUT[12340]$ has a mean of 0.5, and $LUT[11111]$ and $LUT[00110]$ both have means of 1. Some other statistics (only for the non-null table entries) that can be found in the system are shown in Figure 7.2 (right). Note that, as indicated above, other measurements such as the standard deviation and standard error are also available for each N-tuple entry.

Solution	Fitness
$[1, 2, 3, 4, 0]$	1
$[1, 1, 1, 1, 1]$	1
$[0, 0, 1, 1, 0]$	1
$[1, 2, 3, 4, 0]$	0

N-Tuple	Pattern	Mean	Nb. of Eval.
	$[0, *, *, *, *]$	1	1
	$[1, *, *, *, *]$	⅔	3
	$[*, 0, *, *, *]$	1	1
	$[*, 1, *, *, *]$	1	1
	$[*, 2, *, *, *]$	½	2
1-Tuple	$[*, *, 1, *, *]$	1	2
	$[*, *, 3, *, *]$	½	2
	$[*, *, *, 1, *]$	1	2
	$[*, *, *, 4, *]$	½	2
	$[*, *, *, *, 0]$	⅔	3
	$[*, *, *, *, 1]$	1	1
5-Tuple	$[0, 0, 1, 1, 0]$	1	1
	$[1, 1, 1, 1, 1]$	1	1
	$[1, 2, 3, 4, 0]$	½	2

Figure 7.2: Sample points and their corresponding fitness values (left) and some non-null table entries stored in the system (right).

7.3.2   VARIANTS OF ALIENS FOR AGENTS WITH DIFFERENT LOOK-AHEADS

This section shows an example of usage of NTBEA for tuning parameters of the game Aliens. First, a Game Space is created for this game in VGDL, as described above. The resultant search space is described in Table 7.1, including the total search space.

Table 7.1: *Aliens'* parameter set search space

Name	Description	Possible Values	Size
BSPEED	Speed of the aliens' bombs	0.1, 0.2, ..., 1.0	10
ASPEED	Speed of the aliens	0.1, 0.2, ..., 1.0	10
IS_SAM_SINGLE	Are sams (player's bullet) singleton?	True/False	2
ACOOL	Cooldown for aliens' movement	1, 2, 3, 4, 5	5
PTOTAL	Number of aliens to be spawned	10, 15, 20, 25, ..., 60	11
SSPEED	Speed of the avatar's sams	0.1, 0.2, ..., 1.0	10
APROB	Probability of alien dropping a bomb	0.01, 0.05, 0.1, 0.15, ..., 0.75	15
APCOOL	Cooldown for alien spawning portal	1, 2, 3, 4, 5	5
Total search space size		4.95×10^6	

The objective set for this experiment is to find games where an agent A achieves a victory rate as highest as possible, while an agent B loses as many games as possible. Both agents are based on the MCTS agent distributed with the GVGAI framework and the difference between the two resides on the depth of their Monte Carlo simulations. Two depths are used in this study: 20 and 5. It is worth noting that MCTS, using depth 10 and 4 ms for decision time (the same budget has been used for these experiments), achieves 100% victories on the default game of Aliens.

Two different experiments have been conducted.

- Experiment I: agent A (the one whose number of victories must be maximized) has a simulation depth $d_A = 20$, while agent B has a maximum depth of $d_B = 5$.

- Experiment II: $d_A = 5$ and $d_B = 5$.

Note that these experiments are searching for two different objectives: experiment I favors agents with long look-aheads, while experiment II favors those with shorter ones.

Both agents A and B are used to evaluate all points in the search space. Each game is played twice, one per agent. The final outcome ($o_A, o_B \in \{0, 1\}$, where 0 means a loss and 1 a win) and the final score (s_A, s_B) are recorded. The performance of each agent on each game is computed as $V_i = s_1 \times (1 + C \times o_i)$. Note that this expression rewards agents winning the game by a factor C, which multiplies the score obtained by the agent. The fitness of the individual (instantiation of the given game) is defined be the logistic function shown in Equation (7.3):

$$Fitness = \frac{1}{1 + e^{-(V_A - V_B) \times K}}, \qquad (7.3)$$

where K smooths the steepness of the logistic function. Therefore, a fitness value of 0.5 would indicate a similar performance of both agents in the game. A value close to 1 indicates that the agent *A* performs much better than *B* and the contrary is true for values close to 0. Ten repetitions were run for each experiment, lasting for 10,000 generations each, each one with a random seed for the game kept constant during the ran, chosen uniformly at random.

Figure 7.3 show the progression of the fitness as defined in Equation (7.3) for both experiments. Note that these images plot the cumulative average of the fitness from the start until the given generation. Given the noisy nature of the agents and the exploratory component of NTBEA, individual fitness measurement is very noisy and their plot does not offer much information.

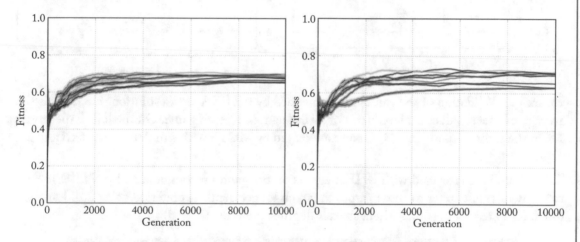

Figure 7.3: NTBEA Fitness progression on Aliens games. On the left, experiment I. On the right, experiment II. All games played by two MCTS agents with simulation depths 5 and 20.

As can be seen, all fitness values average around 0.5 at the start of the runs, in which represents games being explored for which both agents perform similarly. In both experiments (I on the left, II on the right), all runs progressively find games in which fitness is higher (respectively, agent *A* performing better than *B* on the left, vice versa on the right).

Each generation evaluates one point (with no re-sampling) of the search space. It is therefore worth validating that the final games recommended by NTBEA have the desired properties. In order to check this, a validation experiment has been performed by playing 20 times each one of the recommended games (per run, thus 10 in total) and averaging the fitness values achieved. Furthermore, for this validation games are played with different random seeds to the ones used during their evolutionary runs. Figure 7.4 shows that all evolved games achieve a greater than 0.5 fitness, which means that all games are played better by one agent than the other.

Finally, an extra set of games have been played in these recommended games using different agents and an extra set of seeds for the random generators. A rolling horizon evolutionary algorithm (RHEA, see Chapter 3.5) has been used with individual lengths 5 and 20, same as used in the experiments, in order to test that the games evolved do not only have the desired

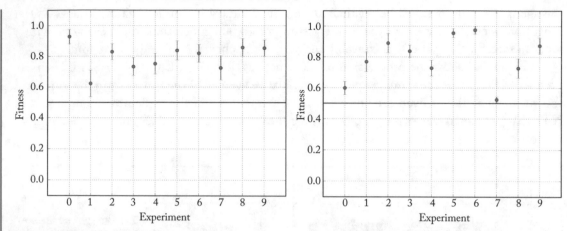

Figure 7.4: Validation of the games recommended by NTBEA after evolution. Each bar is the average of fitness values achieved fruit of 20 evaluations of each game. On the left, Experiment I. On the right, Experiment II. Games are played by MCTS with simulation depths 5 and 20.

Table 7.2: Test performed with RHEA agents on the games recommended by NTBEA. First two row of results refer to those games in which longer depth is preferred ($d = 20$). Last two row refer to results in games where shorter depth is preferred ($d = 5$).

Agent	% Victories	Average Score (std error)	Time Steps (std error)
Results Obtained in Games Evolved for Experiment I			
RHEA (d=20)	100	61.65 (0.92)	101.85 (4.45)
RHEA (d=5)	5	65.45 (1.17)	133.65 (1.79)
Results Obtained in Games Evolved for Experiment II			
RHEA (d=20)	0	29.05 (4.19)	98.8 (8.28)
RHEA (d=5)	100	124.7 (0.84)	344.55 (14.38)

properties when playing with the agents used during evolution. Table 7.2 shows the percentage of victories of each agent, the average score and time steps. As can be seen, the evolved games are robust to the type of agent used to play them and the seeds used.

Examples of Generated Games

Figure 7.5 shows a screenshot of one of the games recommended for Experiment I (favoring longer look-aheads). Table 7.3 shows the final parameters found by NTBEA for this game.

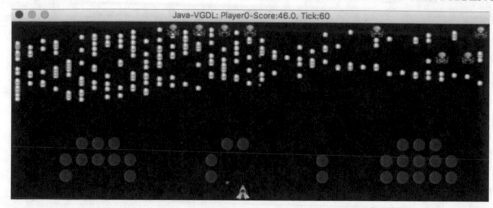

Figure 7.5: Screenshot of the game *Aliens* evolved for the Experiment I, where longer look-aheads are preferred. A video of this game, played by RHEA ($d = 20$), can be found at: `https://www.youtube.com/watch?v=PWHGM_Bd6Jw`.

Table 7.3: Parameters tuned by NTBEA for one of the games of Experiment I

Parameter	Value	Parameter	Value	Parameter	Value	Parameter	Value
BSPEED	0.1	ASPEED	0.8	IS_SAM_SINGLE	True	ACOOL	1
PTOTAL	25	SSPEED	0.6	APROB	0.3	APCOOL	2

As can be seen, the game looks much different than the original one (see Figure 2.2 in Chapter 2). There are many more alien bullets on screen (note parameter BSPEED in Table 7.3 is the minimum non-zero possible speed for the bullets) and with bullet being very close to each other on the y-axis, product of a relatively high probability of shooting ($APROB = 0.3$). This creates a type of games where a *cascade* of bullets is thrown upon the player. The only chance for the player to win the game is to destroy all aliens quickly, before the unavoidable waterfall of bullets reaches the player first. Precision for shooting at the enemies is needed for this, and an algorithm with a longer look-ahead can reach the rewards available by killing aliens in its planning horizon better than an agent with a shorter simulation depth.

Conversely, Figure 7.6 shows a screenshot of a game evolved for Experiment II and Table 7.4 the final parameter values.

As can be seen, the game looks again very different to previous versions of Aliens shown in this book. In this case, the enemies move faster ($ASPEED = 0.1$) and they are more separated from each other ($APCOOL$ higher in this case than in Table 7.4). Bullet speed is also higher than in the previous case, with $BSPEED = 1.0$ as a parameter. The resultant game is much faster paced than the example of Experiment I but in this case, with great skill, it is possible to dodge

Figure 7.6: Screenshot of the game *Aliens* evolved for the Experiment II, where shorter look-aheads are preferred. A video of this game, played by RHEA ($d = 5$), can be found at: `https://www.youtube.com/watch?v=t6usa_f9jig`.

Table 7.4: Parameters tuned by NTBEA for one of the games of Experiment II

Parameter	Value	Parameter	Value	Parameter	Value	Parameter	Value
BSPEED	1.0	ASPEED	1.0	IS_SAM_SINGLE	True	ACOOL	2
PTOTAL	55	SSPEED	0.3	APROB	0.5	APCOOL	4

bullets. This maneuver is easier for an algorithm with shorter look-ahead[2] and allows this type of agent to survive long enough until the enemies reach the simulation horizon, when they can be killed. Longer-sighted agents are typically killed by these bullets as they are less precise on the shorter range.

We found really interesting that the same algorithm, NTBEA, is able to evolve and create new games (with dynamics and relationships between the game parameters that were never thought of before) that respond to opposed objectives in a stable and robust (to noise and agents) way. This result opens the path to a line of research that explores game spaces automatically by using agents to evaluate them.

7.4   MODELING PLAYER EXPERIENCE

In this section we describe our work on using NTBEA to tweak VGDL games with the objective of adjusting the game experience of the players. In particular, the aim is to modify the game parameters so the score progression that their playing agents achieve follows a pre-determined curve.

---

[2]Having shorter depths and the same budget, more certain analysis can be done of the near future events.

7.4.1   DESIGNING THE SEARCH SPACE

In the original VGDL, all sprites that are spawned from *portal* sprites are created at a constant rate during the whole game, in order to enrich the space where interesting games can exist, these portals have been modified to establish a two limits, lower and upper, which determine when the portal is allowed to spawn sprites. These limits are now part of the possible parameters that can be tweaked by any optimization algorithm. It is worth noting that the original version of the games with portals are still valid points in the search space using these limits (concretely, setting the lower limit to 0 and the upper limit to 2,000).

For this work, three games were chosen and parameterized: *Defender*, *Waves* and *Seaquest*. Figure 7.7 shows screenshots of the three games, as in their original implementation. In *Defender*, the player controls an aircraft that aims at destroying some aliens that are bombing a city. The player can shoot missiles but requires ammunition, which is provided via supply packs that fall from the sky. Aliens move from their spawn points horizontally to the left and are harmless to the avatar.

In *Waves*, the player must again try to fight aliens but in this case the objective is survival. Aliens are spawn from the right and move toward the player, located initially on the left part of the screen. Aliens shoot missiles at the player which this must avoid but that can be destroyed by the player's own bullets. When this happens, a shield drops that can be picked up for extra protection. Finally, *Seaquest* is a port of the original game with the same name, in which the player controllers a submarine that must rescue divers spawned at the bottom of the sea. The submarine can stay under water for a certain amount of time before oxygen runs out, before which the player must come to the surface or game is over. Different types of animals (whales, sharks and piranhas) move horizontally and kill the player upon contact.

These games have been chosen due to the possibility of designing large game spaces in them, which can provide multiple different instances of the same game. Tables 7.5, 7.6, and 7.7 describe the search spaces designed for these games (respectively) using the VGDL parameterization model shown in this chapter. All these search spaces reach a size of 10^{10}.

Selected Target Score Functions

The objective of this work is to find, in the search space of these games, particular instances of them where the players are exposed to certain predetermined target score functions. All functions designed for this experiment are positive-definite ($f(x) > 0$ for all $x > 0$). We have defined four functions: *linear* (Equation (7.4), with $m \in \{0.2, 0.4, 1\}$), *shifted sigmoid* (Equation (7.5), with $K_3 \in \{3, 12\}$), *logarithmic* (Equation (7.6), $L = 15$) and *exponential* Equation (7.7). Note that the last two functions require a fast increase in the score rate either at the start or at the end of the games. These functions are plotted in Section 7.4.2 together with the results, for the shake of space:

$$f(x) = mx \tag{7.4}$$

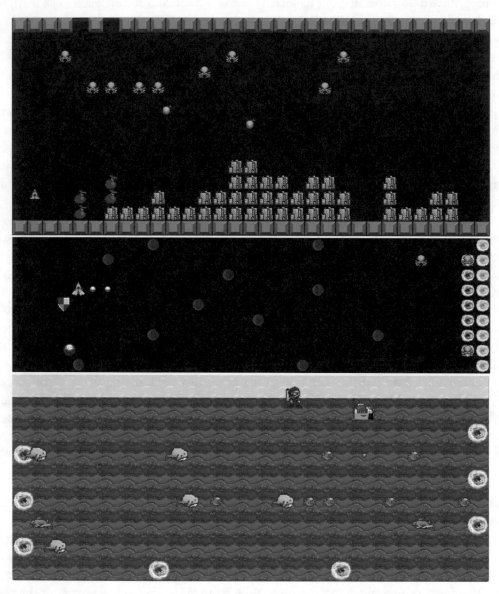

Figure 7.7: Screenshots of the games, from top to bottom: *Defender*, *Waves* and *Seaquest*.

Table 7.5: *Defender*'s parameter set search space

Name	Description	Possible Values	Size
BSPEED	Bomb speed	0.1, 0.3, 0.5, 0.7, 0.9	5
ASPEED	Alien speed	0.2, 0.4, 0.6, 0.8, 1.0	5
SUPSPEED	Supply falling speed	0.05, 0.25, 0.45	3
APROB	Alien's probability to shoot a bomb	0.01, 0.02, 0.03, 0.04, 0.05	5
SLOWPPROB	Slow portal's probability to spawn an alien	0.05, 0.1, 0.15, 0.2, 0.25	5
FASTPPROB	Fast portal's probability to spawn an alien	0.3, 0.5	2
AMPROB	Supply portal's probability to spawn a supply	0.05, 0.15, 0.25	3
ACOOLDOWN	Alien's bomb shooting cooldown	2, 4, 6, 8, 10	5
PCOOLDOWN	Alien portal's cooldown	5, 10, 15, 20	4
AMCOOLDOWN	Supply portal's cooldown	5, 10, 15, 20	4
BLIMIT	Avatar's maximum ammo supply	5, 10, 15, 20	4
ADDSUP	Amount of ammo a supply pack contains	1, 2, 3, 4, 5	5
LOSSCITY	Score lost when a city is destroyed	-4, -3, -2, -1	4
AREWARD	Score gained when an alien is shot	1, 3, 5, 7, 9	5
DELAY	The time step that all portals start spawning	0, 50, 100, 150, 200, 250, 300	7
CLOSE	The time step that all portals stop spawning	350, 400, 450, 500	4
Total search space size		1.08×10^{10}	

Table 7.6: *Waves'* parameter set search space

Name	Description	Possible Values	Size
RSPEED	Rock's speed	0.45, 0.95, 1.45, 1.95, 2.45	5
SSPEED	Avatar missile's speed	0.5, 1.0, 1.5, 2.0	4
LSPEED	Laser's speed	0.1, 0.2, 0.3, 0.4, 0.5	5
ACOOLDOWN	Alien portal's cooldown	2, 6, 10, 14	4
RCOOLDOWN	Rock portal's cooldown	2, 6, 10, 14	4
APROB	Alien portal's probability of alien spawn	0.01, 0.05	2
RPROB	Rock portal's probability to spawn a rock	0.15, 0.2, 0.25, 0.3, 0.35, 0.4	6
PSPEED	Avatar's speed	0.5, 1.0, 1.5	3
ASPEED	Alien's speed	0.05, 0.1, 0.15, 0.2, 0.25, 0.3	6
SLIMIT	Avatar's maximum health point	2, 4, 6, 8, 10	5
ASPROB	Alien's probability to shoot a laser	0.005, 0.01, 0.015, 0.02	4
SPLUS	Avatar's health increase when picking a shield	1, 2, 3, 4, 5	5
APEN	Score lost when avatar collides with alien	-4, -3, -2, -1	4
LASERPEN	Score lost when avatar hit by a laser	-4, -3, -2, -1	4
SREWARD	Score gained when an alien is shot	1, 3, 5, 7, 9	5
DELAY	The time step that all portals start spawning	0, 50, 100, 150, 200, 250, 300	7
CLOSE	The time step that all portals stop spawning	350, 400, 450, 500	4
Total search space size		7.741×10^{10}	

Table 7.7: *Seaquest*'s parameter set search space

Name	Description	Possible Values	Size
SSPEED	Shark's speed	0.05, 0.2, 0.35, 0.5	4
WSPEED	Whale's speed	0.05, 0.2	2
PSPEED	Piranha's speed	0.05, 0.2, 0.35, 0.5	4
DSPEED	Diver's speed	0.1, 0.3, 0.5, 0.7, 0.9	5
SHPROB	Shark portal's probability of shark spawn	0.01, 0.06, 0.11, 0.16	4
WHPROB	Whale portal's probability of whale spawn	0.005, 0.025, 0.045, 0.065, 0.085	5
DHPROB	Normal diver portal's probability of spawn	0.005, 0.015, 0.025, 0.035, 0.045	5
OFDHPROB	Fast diver portal's probability of diver spawn	0.05, 0.07, 0.09	3
WSPROB	Whale's probability to spawn a piranha	0.01, 0.04, 0.07, 0.1	4
HP	Avatar's initial oxygen amount	9, 17, 25, 33	4
MHP	Avatar's maximum oxygen amount	10, 20, 30, 40	4
HPPLUS	Oxygen gained per time step at the surface	1, 2, 3, 4	4
TIMERHPLOSS	Oxygen amount lost per time step underwater	5, 10, 15, 20	4
WHALESCORE	Score increased when a whale is shot	5, 10, 15, 20	4
DCONS	Consecutive tiles a diver can move per step	1, 2, 3	3
CRLIMIT	Max divers the avatar can rescue in one dive	1, 3, 5, 7	4
DELAY	The time step that all portals start spawning	0, 50, 100, 150, 200	5
SHUTHOLE	The time step that all portals stop spawning	200, 250, 300, 350, 400	5
Total search space size		5.892×10^{10}	

$$f(x) = 150 \times \left(\frac{1}{1 + \exp\left(-\frac{x}{30} + K_3\right)} \right) \tag{7.5}$$

$$f(x) = L \log_2 x \tag{7.6}$$

$$f(x) = 2^{\frac{x}{70}}. \tag{7.7}$$

It is important to highlight that these functions are targets, and as such they may be impossible to achieve by a playing agent in the game space defined. However, our objective is to use NTBEA to tune games so the recommended instances provide an experience, as defined by the score trend, that approximate these ideal progression curves as much as possible.

Fitness Calculation

Potential solutions explored by NTBEA are evaluated using the RHEA agent available in the GVGAI framework during evolution. Each game played records the score at every game tick and the final outcome (win or loss). We use a normalized root mean square error (NRMSE) to compute the deviation between the score obtained during the game and the target trend. Let $\hat{s}$ be the vector of scores achieved from timestep 1 to n (the last game tick), and $\hat{y}$ the vector of target scores for a given function. RMSE is calculated as shown in Equation (7.8), which is also the loss function on the target:

$$Loss(\hat{s}, \hat{y}) = NRMSE(\hat{s}, \hat{y}) = \frac{\sqrt{\sum_{i=1}^{n} (\hat{y}_i - \hat{s}_i)^2}}{n(\hat{y}_{max} - \hat{y}_{min})}. \tag{7.8}$$

We then define $1 - Loss(\hat{s}, \hat{y})$ as the fitness function to be maximized by NTBEA.

7.4.2   EVOLVING GAMES FOR PLAYER EXPERIENCE

Experimental Setup

NTBEA has been used to evolve parameters on the three VGDL games described in the previous section: *Defender*, *Waves*, and *Seaquest*, aiming to fit the score progression of an RHEA agent to different variations of the four target functions described above. Furthermore, an MCTS agent has been used to validate the games finally suggested by NTBEA, playing 20 times each one of them. In total, 21 different experiment settings, resulting of testing 7 variations of these target functions, have been tested:

- Linear function, Equation (7.4), $m = 0.2$.

- Linear function, Equation (7.4), $m = 0.4$.

- Linear function, Equation (7.4), $m = 1.0$.

- Sigmoid function, Equation (7.5), $K_3 = 3$ (shifted left).

- Sigmoid function, Equation (7.5), $K_3 = 12$ (shifted right).

- Logarithmic function, Equation (7.6).

- Exponential function, Equation (7.7).

Ten runs have been performed for each one of these settings and outcomes have been averaged to present their results. RHEA was executed using a population size of 20 with an individual length $l = 10$ and mutation rate $1/l$. The C value for the tree policy of the MCTS agent is set to $\sqrt{2}$ and the simulation depth $d = 10$. Both agents count on the same value function to evaluate a state, which follows Equation (7.9). In this scenario, the value would be the score of the game unless the game is over and has been won (1,000) or lost (−1,000). It is important to highlight, thus, that the agents are always aiming to maximize score. Hence, the score trend shown in a game depends mostly on the characteristics of the game itself, with noise introduced by the inherent stochasticity of the games and agents employed.

$$V(s) = \begin{cases} score(s), & \text{otherwise} \\ 1000, & \text{if game won in state } s \\ -1000, & \text{if game lost in state } s. \end{cases} \tag{7.9}$$

Finally, C for the NTBEA bandits is also set to $\sqrt{2}$ and the number of neighbors for NTBEA is established at 100.

Evolving for a Linear Score Progression

Figure 7.8 shows the average fitness trend over 500 generations of NTBEA in the three games studied in this work, when using the linear function $y = 0.2x$ as the target score trend. In this curve (as for all the others), x represents the time step and y the score at time step x. For these plots, the optimal value is 1 (as it minimizes the loss—Equation (7.8)—to 0). The plots represent an average of the 10 runs performed per experimental setting, and the light blue shaded area indicates the standard deviation of the values.

As can be seen, the fitness progression in the three games approximates to 1, which translates to explored points in the search space that represent games for which the score trend that RHEA achieves is close to $y = 0.2x$. Fitness values seem to stabilize at generation 1,000 for all games. In *Seaquest* (7.8-right), convergence was achieved quickly (around generation 200), while *Waves* stabilized a bit later (7.8-center, 500 generations). *Defender* (7.8-left) took more time to reach a stable fitness (albeit with a higher standard deviation), after close to 800 generations.

Another interesting way of analyzing this progression is to plot the actual score trend that the games achieve per generation. As that would be quite difficult to include in a single figure, we have taken average of score trends in consecutive generations. In order to be able to see this

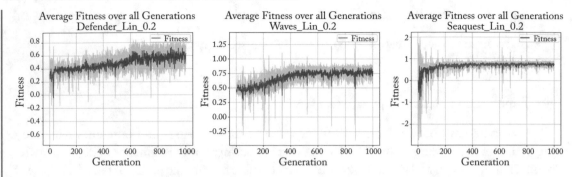

Figure 7.8: Average fitness throughout evolutions for $y = 0.2x$ on the games of this study (left to right: *Defender*, *Waves* and *Seaquest*.

progression, segments are created of N generations each, and an average is plotted for all of them. The value of N is adjusted per game for a better visualization, as different games evolved at different speeds (but N is kept constant though the experiment for all runs). If NTBEA is progressing in the right direction, it is expected that consecutive segments approximate better the target function. Figure 7.9 plots these curves per generation segment for the three games evolved to fit $y = 0.2x$. Note that the red function is the target sought.

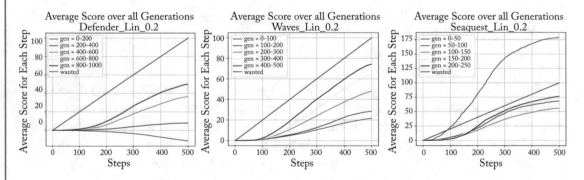

Figure 7.9: Average score trend throughout evolution for the linear function $y = 0.2x$ on the games tested in this study (left to right: *Defender*, *Waves* and *Seaquest*).

As can be observed, the first curves (blue lines) diverge considerably to from the target progression. These are the initial game instances explored by NTBEA, where the algorithm selects points in the search space with little to no information. See, for instance, how in Seaquest (Figure 7.9, right) the first average score trend is above the target by a significant gap. In the other games, the initial score progression lies below the target in *Waves* (Figure 7.9, center), and it is negative (i.e., the player loses points) in *Defender*.

As the landscape model of NTBEA becomes more accurate, the individuals explored show a better performance and the progression curves approximate the target better. In general, the black curves (last generation segment) achieves the smallest error with respect to the target for all games and aimed trends. For all games, NTBEA manages to find parameter sets that are closer to the target. Furthermore, these results are consistent for all linear functions chosen as targets.

As mentioned above, we played the suggested games with MCTS to verify that the games suggested by NTBEA provide the desired target score trends not only for the agent that was used during evolution (RHEA) but also for a different one. This is analogous to the procedure done for Aliens in Section 7.3.2. For doing this validation, the best individual of each run was selected and played 10 times. The score from these games were recorded, averaged for the same parameter set, and plotted along with others in the same evolution configuration. Figure 7.10 shows the score trends of the best individuals found in all evolutionary runs for the three games. To save space, we are only showing the plots for the target function $y = x$ (the figures of the other linear functions are very similar).

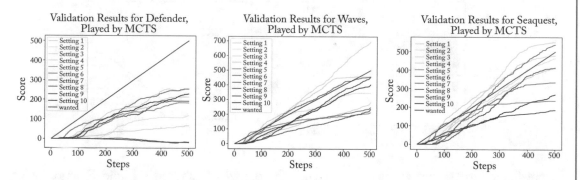

Figure 7.10: Average score trend on validation for $y = x$ on the games of this study (left to right: *Defender*, *Waves* and *Seaquest*).

It is worth highlighting that most of the recommended individuals for *Defender* (Figure 7.10, right) are games for which MCTS achieves a positive score on average (in contrast with the initial negative score trend seen in the first generations during evolution). The target $y = mx$ with $m = 1$ seems to be, however, hard to approximate in this game. in contrast, games suggested for *Waves* and *Seaquest* provide better results, showing score trends in which the MCTS agent approximate the target progression.

Fitting Advanced Score Trends

Figures 7.11 and 7.12 shows the fitness progression and score trends when the left-shifted (respectively, right-sifthed) sigmoid functions are used as targets.

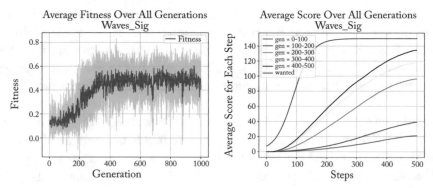

Figure 7.11: Fitness and average score trends for the (left) shifted sigmoid function $y = \frac{150}{1+\exp(-\frac{x}{20}+3)}$ in *Waves*.

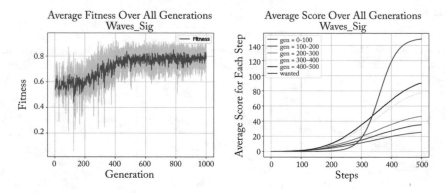

Figure 7.12: Fitness and average score trends for the (right) shifted sigmoid function $y = \frac{150}{1+\exp(-\frac{x}{20}+12)}$ in *Waves*.

Both figures show, on the left, the fitness progression of the NTBEA runs and, on the right, the curve progression as depicted in generation segments. The fitness progression suggests that the left-shifted target is more difficult to adjust that the right-shifted one: the initial fitness of the former is worse and it reaches a constant value of 0.5 (optimum is 1.0), while the latter converges to 0.8.

The left-shifted function requires a game that progresses from providing no score chances to many very early in the game, while the right version of this function shifts this score opportunity change to later in the game. This causes that, in the former case, the average score achieved is higher than in the latter trend, as more game ticks are available to score. This can be clearly seen in both Figures 7.11 and 7.12, right plots. For instance, in the middle game (time step around 250), the score achieved in the left-shifted case is higher than its counterpart. Similarly, the individuals of the right-shifted sigmoid case evolved to provide fewer score opportunities in the first half of the game.

It can be seen how, starting from a similar trend in the first generation segment (blue lines), the left-shifted trend evolves progressions with higher scores, while the right-shifted version stays at low values. The similar starting point is expected (initial random parameters) and it is clear that NTBEA manages to find parameter sets with very differentiated trends in the end, suggesting that this method is general and can adapt to different target functions from the same starting point.

Validation of these games is again performed using MCTS to play the games suggested by NTBEA. Figure 7.13 shows the score trends achieved by MCTS in these games for the left-shifted (left figure) and right-shifted (right) target functions.

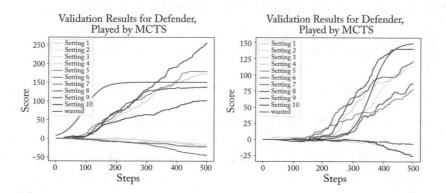

Figure 7.13: Average score trend on validation for *Defender*, shifted sigmoid target functions $y = \frac{150}{1+\exp(-\frac{x}{20}+3)}$.

As can be seen, NTBEA is more successful at evolving games for the right-shifted function. Acknowledging some evolved games fail dramatically at providing the desired score trend in both cases (horizontal blue lines in Figure 7.13), most of the games evolved for the right-shifted progression adjust better to the desired progression than their counterparts of the left-shifted curve. For this one, the agents achieve (in the best case) close to linear score trends. This suggests again that the left-shifted sigmoid target is harder to approximate, and this actually is understandable: the game must start with no score opportunities to quickly provide more, but stabilizing again in less than 200 game steps at a maximum score. Again, it is worth remembering that the target trends may be not achievable by the agents in the game spaces available—they must be understood as guides for evolution.

The results for the logarithmic and exponential target functions are very similar to the ones obtained for the sigmoid ones. Again, we plot the validation results in the game *Defender*, which has shown to be the most challenging game of the three. Figure 7.14 shows the results of average

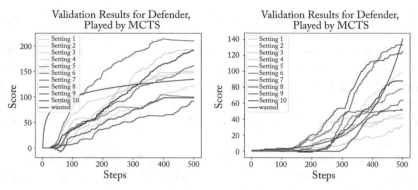

Figure 7.14: Average score trend on validations for the game *Defender*, logarithm and exponential target functions ($y = 15\log_2(x)$ and $y = 2^{\frac{x}{70}}$, respectively).

score for *Defender*, targeting the logarithmic (left) and exponential (right) score progression functions.

In this case, the logarithmic trend again poses more problems to find games that adhere to the target curves than the exponential one. The reason for this is the quick and sudden requirement for score in the first steps of the game (it requires to achieve 35 points in the first 5 game steps). However, it is clear that the evolved games behave similar under the same MCTS agent that plays them: the trends shown in both images from Figure 7.14 are noticeably different.

We observed the values of the parameters evolved by NTBEA in *Defender*, the game that has shown to be the hardest of the three studied here. In linear functions, a higher slope ($m = 1$) in the trend tended to produce games with a higher supply limit, slower bombing speed and alien spawn probability. Also, the alien spawn portal stops creating enemies earlier on higher slopes than in the others.

Observing values evolved by NTBEA for the advanced target functions in *Defender*, we can see that the supply amount, alien movement speed and alien spawning rate are higher in the right-shifted sigmoid function than in the left-shifted one. Portals used for spawn aliens open and closed later in the exponential version, providing a slower supply speed and faster alien movement.

Examples of the recommended *Defender* games can be found in an online video.[3] These show some games evolved for each one of the target function families (linear, sigmoids, logarithmic and exponential), where the differences can be easily observed for each setting.

---

[3]https://youtu.be/GADQLe2TiqI

7.5   EXERCISES

The GVGAI Framework is available in a Github repository.[4] Use the release 2.3[5] in order to run the same version presented in this chapter. The NTBEA algorithm can be found at its Github repository[6] and you can use the checkpoint[7] for a stable version of the code. Additionally, you can also find a Python version of NTBEA at another Github repository.[8,9]

7.5.1   PARAMETERIZING VGDL GAMES

VGDL game spaces can be found in the `examples/gameDesign` folder of the GVGAI repository.

Editing a Game Space

You can play-test parameterized games running the class `tracks.gameDesign.TestGameSpace`. Take one of the games in `examples/gameDesign` and open the VGDL description. Study how modifying the values of the parameters change the game when you play it.

Parameterize a VGDL Game

Take any game from `examples/gridphysics`, `examples/contphysics` or `examples/2player` and copy it to `examples/gameDesign`. Then, parameterize this game considering the possible values the parameters may have. Remember to do the following.

- Change the game type (first line) from `BasicGame` to `GameSpace`.

- Add the set `ParameterSpace`, in which you will define all parameters that you want to add to the game.

- For each boolean, integer or floating point value available in the `SpriteSet`, `InteractionSet` and `TerminationSet`, you may add a new parameter to the game space. Think about the impact that it has in the game, and define feasible minimum and maximum values for the parameter, as well as possible intermediate values defining an increment step.

7.5.2   OPTIMIZE VGDL GAMES

Use NTBEA to tweak a parameterized VGDL game so it behaves in a determined manner. Examples are as follows.

---

[4]https://github.com/GAIGResearch/GVGAI
[5]https://github.com/GAIGResearch/GVGAI/releases/tag/2.3
[6]https://github.com/SimonLucas/ntbea
[7]https://github.com/SimonLucas/ntbea/commit/2bb178d5ea57c4219b12a49aea32d59ed596612a
[8]https://github.com/bam4d/NTBEA
[9]These exercises are also available at this book's website: https://gaigresearch.github.io/gvgaibook/.

- The game should be easier to complete for high skilled agents than low skilled ones.

- The game should provide a determined player experience. Score trends was studied in Kunanusont, Lucas, and Pérez Liébana [2018], but other options are possible: feel of danger, number of different solutions, pace, etc.

- Evolve game parameters so they provide a specific *bot* experience. For this, you may look at the agent experience features described in Section 4.4. The objective game configuration should aim at providing certain values for these features (i.e., reducing the decisiveness of the agent).

You can also tweak the NTBEA parameters and observe if in any of the above scenarios the optimization converges faster to a desirable solution.

CHAPTER 8

GVGAI without VGDL

Simon M. Lucas

8.1 INTRODUCTION

In this chapter, we outline the motivation and main principles behind writing GVGAI compatible games in a high-level language instead of using VGDL. The specific examples described below are all written in Java, but the principles apply equally well to other efficient high-level languages. Java, including its graphics APIs, runs on Windows, macOS and Linux without requiring any additional libraries or plugins. We are currently evaluating the use of the new Kotlin[1] programming language as an even better alternative to Java. Kotlin has many language features that lead to clearer and more concise code while being fully inter-operable with the Java platform (Kotlin code compiles down to Java Virtual Machine—JVM—code). Kotlin also has other advantages, such as strong support for writing Domain Specific Languages [Jemerov and Isakova, 2017] (which would support flexible languages for describing particular game variants, or the development of new versions of VGDL). Furthermore, Kotlin can also be transpiled to JavaScript and then run in a web browser, if all the code is pure Kotlin and does not call any Java or native libraries. While it used to be possible to run Java Applets in a web browser, in recent years this has become increasingly difficult with most web browser security managers preventing it.

First the motivation. VGDL is a good choice for expressing relatively simple games where the interactions of the game objects and the effects they trigger can be expressed via condition action rules involving rudimentary calculations. When these conditions hold, games can be authored quickly with concise code that is easy to understand and describes the core elements of the game while avoiding any distractions.

However, the restrictive nature of VGDL makes it relatively hard to implement games with complex and strategically deep game-play. Furthermore, existing VGDL game engines run much more slowly than a carefully implemented game in an efficient high-level language such as Java, C++ or C#. To give an idea of the scale of the difference, using a high-level language instead of VGDL may make the game 100 times faster. The increase in speed is important not only to run the experiments more quickly, but also to enable SFP methods to be run in real-time with larger simulation budgets. This enables the agents to behave in more intelligent and more interesting ways and provide a better estimation of the skill-depth of the game.

[1]https://kotlinlang.org/

Another factor is programmer preference: despite the elegance of VGDL, many programmers would still prefer to author games in a more conventional high-level language.

GVGAI games are designed primarily to test the ability of AI agents to play them, in the case of most of the GVGAI tracks, and to design content for them in the level generation track. In judging the quality of the AI play and the generated content however, it is also important to make them human-playable. This enables us to compare the strength of the AI with human play and also allows players to evaluate the generated games and content. Although recent efforts have extended VGDL in a number of interesting ways, such as adding continuous physics, the language still imposes some hurdles to be overcome both in terms of the game rules and the modes of interaction. For example, by default a spaceship avatar can either move or shoot at any particular time, but a human player would often wish to do both, and currently GVGAI has no support for any point and click input devices such as a mouse or trackpad.

VGDL-based games are designed around a fixed number of levels to test the generalization of the agents. An even better alternative is to allow random or other procedural level generation methods, which is easier to achieve in a high-level language.

Note that GVGAI aims for really general game AI that can be thoroughly tested on a number of dimensions, and has the potential to go far beyond the Atari 2600 games used in the Atari Learning Environment [Bellemare et al., 2013] both in terms of variability of challenge and depth of challenge.

The benefits of using high-level languages for GVGAI are therefore clear. For the game designer or AI researcher, the advantage of making a game GVGAI compatible is also significant: for a little extra effort it enables a great many agents to be tested on or to play-test the game. Following the efforts of Torrado et al. [2018], GVGAI games are now available in OpenAI Gym [Brockman et al., 2016], something that would also be possible for the example games in this chapter, and future games that conform to the same interfaces.

Finally, each game we implement should really be thought of as a class of games with many variable parameters to control the gameplay. This helps to justify the extra effort involved in implementing a Java game since it may offer a wide range of test cases, with different parameter settings leading to very different gameplay.

8.2   IMPLEMENTATION PRINCIPLES

In this section we outline the main considerations when implementing these games. The steps taken below are not difficult, and help lead to well-designed software. The approach is most easily taken when implementing a game from scratch—retrofitting these design principles to an existing implementation is significantly harder.

8.2.1   COPYING THE GAME STATE

Among all the GVGAI tracks, the planning tracks have been the most popular, especially the single-player planning track. The main requirements for planning track compatibility are to be able to copy the current game state, and to advance it much faster than real-time. Of these, it is the copyability that requires the greatest programming discipline to achieve. For a well-structured object-oriented design, the game state will often comprise an object graph consisting of many objects of many classes. This does not present any great difficulty, but it does mean that care must be taken to make a deep copy of the game state. Also, static or global variables must be avoided. Circular references are also best avoided since they cause problems for serialising the game state to JSON, as JSON does not support circular references.

There are two approaches to copying the game state: either a generic serializer/deserializer such as GSON[2] or WOX[3] can be used, although note that GSON requires that the game state object graph must not contain any circular references (which is a limitation of the JSON rather than the GSON library *per se*). Note that serialising the game state is also useful in order to save particular states for future reference and also to allow agents to interact with the game via network connections. The generic options allow the copy function to be implemented in a couple of lines of code, but they are not very flexible. They are inherently slower than bespoke copy methods, as they use run-time reflection operators which are slower than regular method calls.

Therefore, it is often preferable to write bespoke game-state copy methods. In addition to providing greater speed they also have the flexibility to only make a shallow copy of large immutable objects if desired. For example, a game may contain a large geographical map which does not change during the course of the game. In such a case making a shallow copy (i.e., only the object reference to the map is copied) is sufficient and faster, and also saves on memory allocation/garbage collection.[4]

When implementing a game, there are often many parameters that significantly affect the gameplay. A particular combination of parameter settings defines one game in a potentially vast space of possible games. This game-space view has two benefits. We can test agents more thoroughly by selecting multiple games from the space to be sure that they are not over-fitting to a particular instance of the game. Also, we can use optimization algorithms to tune the game to meet particular objectives. For maximum flexibility we recommend storing every game parameter in a single object, defined in a data-bound class. For example, suppose we have a PlanetWars game, we then define a class called PlanetWarsParams. Each object of this class defines a particular combination of game parameters, and hence a particular game among the set of possible games in this space.

When copying the game state, a deep copy must also be made of the parameters object. This enables experimenting with different game-versions concurrently. An interesting use-case is as follows: the SFP agent can be given a false FM, i.e., one in which the parameters have been changed compared to the ones used in the game under test. Experimenting with false

---

[2]https://github.com/google/gson

[3]http://woxserializer.sourceforge.net/

[4]The potential value of writing bespoke copy methods is clear when using Java, but may be unnecessary when using Kotlin. Kotlin has special data classes which provide default copy methods, and it also has immutable types.

forward models provides valuable insights into how bad a model can be before it becomes useless. Interestingly, initial tests show the model can be highly inaccurate while still being useful.

8.2.2   ADVANCING THE GAME STATE

SFP algorithms typically make many more calls to the next state function than they do to the copy method. For example, rolling horizon evolution or Monte Carlo tree search with a sequence length (roll-out depth) of L will make a factor of L more next state calls than game state copies. For video games, values of L typically varies between 5 and 500, so this means care must be taken to implement an efficient next state function.

Beyond following normal good programming practice, the main ways to do this are as follows.

- Use efficient data structures for detecting object collisions.

- Pre-compute and cache data that will be regularly needed.

- In the object-update loop, it may be unnecessary or even undesirable to update every object on every game tick.

The simplest and most efficient data structure for detecting object collisions is an array which can be arranged as a 2D grid for 2D games and offers constant-time access (the time to access a grid element does not depend on the size of the grid or the number of objects in it). For 3D games Binary Space Partition trees are a common choice, offering access which has logarithmic cost with respect to the number of entries.

An example of pre-computation may be found in the extended version of PlanetWars we implemented [Lucas, 2018]. Each planet exerts gravitational pull on the ships in transit, but this is pre-computed as a vector field and stored in a 2D array, speeding the next state function by a factor of the number of planets N when there are many ships in transit.

Regarding object-updates, some games even use skipped updates to improve the game play. For example, the original Space Invaders by Taito (1979) would only move a single alien in each game tick. This led to two emergent and interesting effects. One is that the aliens move in a visually appealing dog-legged fashion, only lining up occasionally. The other is that the remaining aliens speed up as their compatriots are destroyed. The increase in speed, an effect which is most extreme when going from two to one remaining alien, makes the game much more fun to play and provides a natural progression in difficulty within each level.

8.3   INTERFACING

The original GVGAI interface as seen by the AI agents seems over-complex. For the new Java games, we designed a simpler interface which is still suitable for the SFP agents and is exactly the same for 1- and 2-player games (or even n-player games). This is called AbstractGameState and is listed in Table 8.1. The `next` method takes an array of int as an argument where the ith element is the action for the ith player. The other methods are self-explanatory, although the `nActions` method assumes that each player will have an equal number of actions available in each state, something which is clearly not true in general and will be changed in future versions.

All the games presented in this chapter implement the `AbstractGameState` interface. All agents should implement `SimplePlayerInterface`. Note that the `getAction` method includes the id of the player: this is to support multi-player games (such as Planet Wars) within the same interface. The `reset` method is useful for Rolling Horizon Evolution agents that would normally use a shift-buffer to retain the current best plan between calls to `next`.

Table 8.1: The `AbstractGameState` interface

```
interface AbstractGameState {
    AbstractGameState copy();
    AbstractGameState next(int[] actions);
    int nActions();
    double getScore();
    boolean isTerminal();
}
```

Table 8.2: `SimplePlayerInterface`: an interface which all compatible agents implement. The `reset` method is included because some agents otherwise retain important information between calls to `getAction`.

```
interface SimplePlayerInterface {
    int getAction(AbstractGameState gameState, int playerId);
    SimplePlayerInterface reset();
}
```

8.3.1 RUNNING GVGAI AGENTS ON THE JAVA GAMES

The main purpose of GVGAI-Java is to expand and enrich the set of GVGAI games. To meet this requirement, we implemented a wrapper class that maps each AbstractGameState method to its equivalent standard GVGAI method. There are some limitations: methods such as `getAvatar` are not available in this cut down version: calling a non-implemented method throws a runtime exception. This means that many heuristic methods will not work, and unfortunately excludes some leading GVGAI planning agents such as YOLO-Bot.

For some games this is almost unavoidable: games such as Planet Wars do not have an avatar. While it may be possible to modify a game to include an avatar in a meaningful way, the result would be a different game (possibly an interesting one).

For more typical arcade games, the solution is to extend the `AbstractGameState` interface with a new interface called (say) `ExtendedAbstractGameState`. This will have all the required GVGAI method signatures. All compatible games would then implement the extended interface.

Having implemented the wrapper class, we can now run the existing GVGAI agents on an extended set of games. Just to be clear, at the time of writing this only works for the agents which are solely reward based and do not use heuristics based on observable game features. As explained above, the extension to do this is straightforward for typical arcade games but has not yet been done.

8.3.2   RUNNING NEW AGENTS ON THE VGDL GAMES

When developing agents for the new Java-based games it is also desirable to benchmark them on existing VGDL games. One reason is that the Java games are much faster and allow more effective tuning of the agent parameters. It is therefore interesting to test how well the newly tuned agents perform on the VGDL games.

The approach is simply the opposite of the previous one: we now take a standard VGDL game and provide a generic wrapper class that implements the AbstractGameState interface. Hence, with this one class we enable our new agents to play any of the VGDL games.

8.4   SAMPLE JAVA GAMES

This section outlines some Java games that fit the model proposed here.

8.4.1   ASTEROIDS

Asteroids, released to great acclaim in 1979, is one of the classic video games of all time, and Atari's most profitable.[5]

The challenge for players is to avoid being hit by asteroids, while aiming at them accurately. There are three sizes of asteroids: large, medium and small. Each screen starts with a number of large rocks: as each one is hit either by the player, or by a missile fired by an enemy flying saucer, it splits in to a smaller one: large rocks split into two medium ones, medium into two small ones, small ones disappear. There is also a score progression, with the score per rock increasing as the size decreases. The arcade version features two types of flying that appear at various intervals to engage the player in combat. The version implemented here does not include the flying saucers, but still provides an interesting challenge.

A key strategy is to control the number of asteroids on the screen by shooting one large one at a time, then picking off a single medium rock and each small one it gives rise to, before breaking another large rock. Shooting rocks randomly makes the game very hard, with a potentially great many rocks for the player to avoid.

---

[5]https://en.wikipedia.org/wiki/Asteroids_(video_game)

Figure 8.1 gives a screenshot of the game screen. Pink lines illustrate the simulations of the rolling horizon evolution agent using a sequence length of 100 and controlling the spaceship. These are shown for illustrative purposes only, and ignore the fact that each rollout involves other changes to the game state such as firing missiles and the rocks moving and splitting.

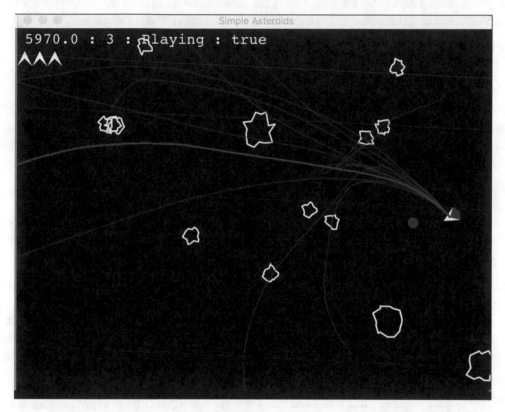

Figure 8.1: Java version of Asteroids, compatible with GVGAI. The game runs at a high speed but includes legacy code which does not implement the recommended approach to parameterization.

Comparison with VGDL Version

Although VGDL has been extended to allow continuous physics and includes a version of Asteroids, the main advantages of the version presented here, enabled by the use of a high-level language are:

- much faster (Java version runs at around 600,000 game ticks per second);

- easy to generate random games (in fact every game starts with different random rock positions and velocities);

- easy to procedurally generate rocks, and rotate them as they move. This is mostly cosmetic, though for large jagged rocks it can affect the game play;

- overlays to show a key aspect of the rollouts. This helps provide insight for a researcher when tuning an agent: for example, if the projections all end in the same or very similar positions then the rollouts show insufficient exploration of the state space;[6] and

- better controls for a human player (can thrust while firing and rotating).

8.4.2   FAST PLANET WARS

Planet Wars (also called Galcon) is a relatively simple real-time strategy (RTS) game that has many free versions available for iOS and Android. The Java version shown in Figure 8.2 is described in more detail by Lucas [2018]. The game involves many features that would lie beyond the current capabilities of VGDL, such as the following.

- Gravity field (shown by vector lines on the map) causes ships to follow curved trajectories.

- Allows plug and play of different actuators (i.e., different ways for a human player or AI agent to control the game). The simplest actuator involves clicking a source then destination planet to transfer ships between.

- An alternative actuator uses the spin of the planets to add an additional skill to the game. This involves clicking and holding on the source planet. While the planet is selected, ships are loaded on to a transit vessel. When the planet is de-selected (e.g., the mouse button is released) then the transit departs in the current direction it is facing. Combined with the gravity field, this adds a significant level of skill to the act of transferring ships between the intended planets.

- The implementation has also been designed for speed, and runs at more than 800,000 game ticks per second for typical game setups.

The game includes many parameter settings as described in Lucas [2018] including the number of planets, the ship growth rates, the range of planet radii, the size of the map, the ship launch speed and the strength of gravity. All of these parameters have significant effects on the gameplay.

8.4.3   CAVE SWING

Cave Swing is a simple side-scrolling casual game where the objective is to swing through the cave to reach a destination blue zone at the rightmost end, while avoiding all the pink boundaries. It is a one-touch game suitable for a mobile platform. The avatar is the blue square. Pressing the

---

[6]Perhaps counter-intuitively, uniform random rollouts may exhibit exactly this problem. When the ship is traveling forward at high speed it takes a more focused effort to turn it around than is afforded by uniform random rollouts.

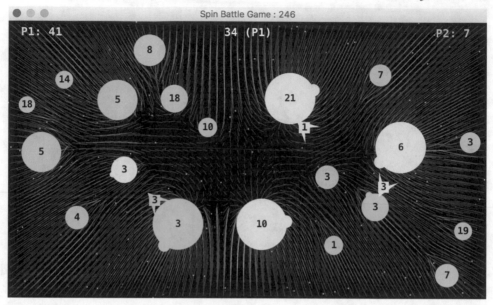

Figure 8.2: A fast implementation of Planet Wars with spinning planets and a gravity field, both of which have significant effects on the game play. The game is compatible with GVGAI and adopts the recommended approach to parameterizing the game.

space bar attaches an elastic rope to the nearest anchor point, releasing it disconnects the rope. Both connection and disconnection happen with immediate effect. Anchor points are shown as grey disks, apart from the nearest one which is shown as a yellow disk. Points are scored for progressing to the right and upward, but each game tick spent loses points. There is also a penalty for crashing and a bonus for completing the game within the allotted time. Hence the aim of the game is to travel to to the right as quickly as possible and finish as high up the blue wall as possible.

The avatar falls under the force of gravity, but this can be more than counteracted by the tension in the elastic rope when attached. The tension force is calculated using Hooke's law using the assumption that the natural length of the rope is zero, so the tension is proportional to the length of the rope.

Cave Swing follows the recommended approach of having all the game parameters bundled into a single object (of type `CaveSwingParams`) (Figure 8.3). Significant parameters include: the size of the cave (width and height), the gravitational force (specified as a 2D vector to allow horizontal as well as vertical force), Hooke's constant, the number of anchor points, the height of the anchor points and various score-related factors.

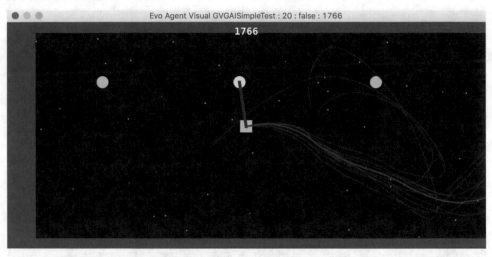

Figure 8.3: The simple but fun casual game we call *Cave Swing*. This runs at high speed and adopts our preferred approach to bundling all parameters into a `CaveSwingParams` object.

8.4.4 SPEED

Table 8.3 shows the timing results for the three games outlined in this chapter running single-threaded on an iMac with core m5 processor. For comparison we also include two typical VGDL games running on the same machine. The speed-up achieved with the Java games is clearly significant.

Table 8.3: Speed of the `next` method in units of thousands of operations per second (iMac with 3.4 GHz Intel Core i5 CPU running single-threaded)

Game	kop/s (1)
Asteroids	600
Planet Wars	870
Cave Swing	4,800
Aliens (VGDL)	65
SeaQuest (VGDL)	61

8.5   CONCLUSIONS

In this chapter, we outlined three games that are compatible with GVGAI but written in Java rather than VGDL to enable faster operation, more flexible gameplay and additional bespoke visualizations (the pink overlaid lines showing expected agent trajectories for each rollout). Authoring games in a high-level language offers some clear advantages and disadvantages compared to using VGDL, but we believe the advantages are significant enough to make this an important future direction for general game AI research. The high speed makes the approach especially well suited to SFP methods such as MCTS and RHE.

Each of the games presented in this chapter has a number of parameters to vary, many of which significantly affect the gameplay. Each game is therefore one of many possible games and we encourage testing agents on many instances of the game in order to provide more robust results. The game parameters can also be optimized to achieve particular aims, such as being particularly hard or easy for certain types of agent, or to maximize skill-depth. Due to the speed of the Java games and the sample efficiency of the N-tuple bandit evolutionary algorithm (NT-BEA) [Lucas, Liu, and Pérez Liébana, 2018a], games can now be optimized in less than 10 s in some cases.

8.6   EXERCISES

All the code is available in a Github repository.[7] Use the checkpoint[8] in order to run the same versions presented in this chapter.[9]

8.6.1   RUNNING THE DEMOS

The first exercise is to download the code and run the following demos. We recommend doing this in a Java IDE such as Intellij IDEA or Eclipse. This enables the code to be easily modified and re-run. Running the demos below will should take a around 5 min. If you spend time adjusting the agent's parameters then this could take 30 min or more.

Each demo shows the same RHE agent playing the game. Note how well the agent is able to play each one without any modification.

Asteroids
Running the following class will run the default RHEA agent. `asteroids.RunAsteroidsDemo`

Planet Wars
Run: `spinbattle.actuator.SourceTargetActuatorTest`. It will play the standard RHEA agent (Player 1, yellow) against a hand-coded heuristic agent (PLayer 2, Blue). You can also play against any of the agents. For example, running `spinbattle.test.HumanInterfaceTest` pits you against a heuristic player. You play as they yellow player. To play the game you click a source planet (must be owned by you, hence yellow) followed by the target planet you wish to invade

---

[7]`https://github.com/ljialin/SimpleAsteroids`
[8]`https://github.com/ljialin/SimpleAsteroids/commit/6ff5e7a8881be1e6d491edb527dd3bdcd9489b25`
[9]These exercises are also available at this book's website: `https://gaigresearch.github.io/gvgaibook/`.

(if blue or grey) or support if already owned by you (yellow). The interface is a bit basic, in that clicking the source planet has no discernible effect until the target planet is subsequently clicked.

Cave Swing

Run: `caveswing.test.EvoAgentVisTest` to test the default RHEA agent. To play the game, run: `caveswing.test.KeyPlayerTest`.

8.6.2   RHEA EXERCISE

All the above demos use the default RHEA agent with the same parameters. Change each one from its default value and observe the effects.

- `useShiftBuffer`: the default is `true`. What is the effect of changing it to `false`?

- `sequenceLength`: the default is 100. What are the effects of much lower values such as 10 on each game?

- `nEvals`: the default is 20. Explore how low can you go before performance significantly degrades?

8.6.3   AGENT COMPARISON EXERCISE

MCTS is much more widely used than RHEA, although in our experience RHEA is simpler to implement and often provides better performance. The exercise is to make some of your own comparisons. Start by running this test on planet wars that runs a round-robin league: `spinbattle.league.RoundRobinLeagueTest`. If using the exact checkpoint given you will see that some of the possible players have been commented out. Include the MCTS player and see how it compares to the RHEA player. Set the total number of games high enough to obtain statistically significant results. This will typically take between 5 min to 1 h depending on the number of games chosen, the number of players in the league and the simulation budget allowed for each of the SFP players.

8.6.4   GAME TUNING EXERCISE

We can put the speed to good use by running some efficient tuning of the game parameters. A starting point for exploring this is provided. Begin by studying and running: `hyperopt.TuneCaveSwingParams`. This takes around 10 s for one complete optimization run. It uses the NTBEA to perform an efficient optimization of the game's parameters. Observe the output of the run, which provides detailed statistics on the effects of various parameter combinations.

Within the output, you will find the optimized solution encoded as an array of int, together with a listing of each parameter value. Try copying these values back in to the `KeyPlayerTest` code (see examples in method `getSearchSpaceParams` of `KeyPlayerTest` and ensure that the call to it on line 21 is uncommented) and play through them.

By default the optimization is set up to maximize the score difference between a mediocre RHEA agent and a smart RHEA agent. This often produces fiendishly difficult levels for human play. As an exercise, change the "dumb" player in `caveswing.design.CaveSwingFitnessSpace` to be a random player (lines 140–142), and observe the effects on the evolved games. You may find them a bit easier to play now!

CHAPTER 9

GVGAI: What's Next?

Diego Pérez Liébana

In the previous chapters of this book we have described the GVGAI framework and competitions, and also some of the most interesting lines of research that have used this benchmark to tackle GVGAI problems. This is, however, a non-exclusive list, as a recent survey shows [Pérez Liébana et al., 2018b]. We have also seen how this framework has been employed in education, with higher education institutions using GVGAI for undergraduate and postgraduate assignments, final year, master and Ph.D. projects. The more technical chapters also provide a list of exercises that can serve as project ideas to take this further.

Research-wise, most of the attention of the GVGAI framework has been devoted to the single and two-player version of the framework. Section 4.3 in this book outlines the main outstanding problems of this challenge: no studies have achieved better than approximately 50% victory rate with any technique when applying it to sets of 10 or more games. Improvements on methods proposed typically boost performance in a subset of games, but not in all. Therefore, one of the most challenging directions at the moment is to increase play performance in a broad variety of games.

Given that, in most cases adding features that are GVGAI focused leads to uneven improvements, it is sensible to think that a potential way forward is to work on feature extraction that is independent from the framework. An example of this is the work described in Chapter 4, where features are based on the agent's experience instead of the game. Recent advances in deep convolutional neural networks could also be a fruitful line of future work. We have described other possible improvements for single-player planning in Section 4.3.

In addition, the two-player planning problem adds an extra complexity on the GVGAI challenge. Not only the agents need to adapt to any game is given, but also they need to compete or collaborate with another agent they know nothing about. The current literature shows no study that tries to identify the behavior of the opponent, with regards to the other player being cooperative or competitive. This type of analysis taps on research on game theory and opponent modeling, being a field of study on its own right. So far, the most involved studies on GVGAI have tried to model the action distribution observed by the other player, using this model for the opponent actions in the forward model [Gonzalez-Castro and Pérez Liébana, 2017]. Investigation on more complex opponent models is another line of future work that could yield better results in this problem.

Using GVGAI for AI-assisted game design, as seen in Chapter 7, is another area of fruitful future research. Using agents to automatically test and debug games have attracted the interest in researchers in the last years, for instance using RAPP, where how good a game is gets evaluated by a measure of the agent performance in them. In a general setting, this can be enhanced using factors like the amount of level being explored by agent, the decisiveness or convergence rate of the agent [Volz et al., 2017] or the entropy of the moves made.

The GVGAI framework can also be subject to further enhancements. For instance, one of the potential areas of future work is to expand the range of games that VGDL can build. At present, there is no support for games with large state spaces, games that are not avatar-centric (i.e., like Tetris or Candy Crush), board games, and card games. Although the concepts described in Chapter 8 can alleviate this, there is a scientific interest in making VGDL more complete. Adding more games can also be achieved by making use of the rule generation tracks (see Chapter 6) and the game parameterization tool (Chapter 7).

This can be complemented with adding an integration with other systems. Different general frameworks like OpenAI Gym [Brockman et al., 2016], ALE [Bellemare et al., 2013] or Microsoft Malmö [Johnson et al., 2016] already count on a great number of games (single, multi-player, model free and model based). Interfacing with these systems would increment the number of available games which all GVGAI agents could play via a common API.

Multi-agent games have also drawn people's attention, for instance, real-time strategy games (e.g., StarCraft) and board games (e.g., Mahjong). The study of multi-agent GVGAI is a fruitful research topic. Atari games can also be extended to multi-agent games. In particular, the Pac-Man can be seen as a multi-agent game and related competitions have been held since 2011 [Philipp Rohlfshagen and Lucas, 2017]. The most recent Ms. Pac-Man vs. Ghost Team Competition [Williams, Pérez Liébana, and Lucas, 2016], which included partial observability, was held at CIG in 2016. Nevertheless, a more general multi-agent track is favorable. Another interesting possibility is to enhance these games by providing a larger range of available actions, either simultaneously (as in Atari games or others were game-pad controllers are used) or in a continuous action space (like joystick or steering wheel controllers in racing games).

The upcoming years could also see the development of the GVGAI competition, potentially adding a few more tracks set up as challenges. The following is a non-exclusive list of possibilities for future tracks.

- Game generation track: the goal of this track would be to provide systems that automatically generate completely new games. Concrete instalments of this track could provide specific themes, or a sub-space of rules to narrow down the space of possibilities. Ideally, generators would create games from scratch, also this challenge is open to modification on existing games as a stepping stone toward full-game generation.

- Automatic game design: there is scope for running a competition track on game tuning and play-testing (player experience, game feeling, fun, etc.). This can be in combination to other generation tracks, such as the level, rule or game generation challenge mentioned in the previous point.

- Multi-Agent GVGAI: in all VGDL games used in the GVGAI competition, the player controls only one agent. An interesting extension of this challenge would be a setting in which the controller decides the moves of a set of agents, for instance in games with more strategic depth like Hero Academy or Starcraft.

- Multi-player GVGAI: scaling up the two-player GVGAI track to accommodate multiple players is another potential future track for the competition. This would allow for an extended research framework where agents must account for the actions of more than one external agent. Games like Bomberman (where four players play with an avatar each) or Civilization (N players controlling M units each, for a multi-agent multi-player setting) would take part in this track.

- Turing Test GVGAI: the idea of applying a turing test, which in this case determines if a game is being played by a human or by a player, is not new [Lehman and Miikkulainen, 2015]. Adding such a track to GVGAI would allow the community to investigate the general features that make an agent play like a human in any game.

Apart from the framework and the competition, a point of future work is the website that holds the competition.[1] The feedback provided to competition participants can be enhanced with more data analysis on results, visualizations, action entropy, exploration and other features that can be extracted from gameplay, providing a more flexible support for gameplay metric logging. Including the possibility of playing (and replaying) games on the web browser and being able to analyze these features in real-time can help analyze the performance of the agents to build AI improvements.

As a final word, and reflecting on the multiple lines of future work available, we can safely say that GVGAI is very much alive. There is plenty of potential for further development and ideas to take this research further. The framework, being open source, is available for anyone to provide improvements and propose new challenges. We have discussed a few here, and provided lists of extensions and exercises on each chapter, but there is no limit in terms of what other things can be done.

---

[1]www.gvgai.net

Bibliography

Anderson, D., Stephenson, M., Togelius, J., Salge, C., Levine, J., and Renz, J. 2018. Deceptive games. *ArXiv Preprint ArXiv:1802.00048*. DOI: 10.1007/978-3-319-77538-8_26 89

Apeldoorn, D. and Kern-Isberner, G. 2017. An agent-based learning approach for finding and exploiting heuristics in unknown environments. In *COMMONSENSE*. 93

Auer, P., Cesa-Bianchi, N., and Fischer, P. 2002. Finite-time analysis of the multiarmed bandit problem. *Machine Learning*, 47(2):235–256. DOI: 10.1109/cec.2016.7744106 33

Beaupre, S., Wiles, T., Briggs, S., and Smith, G. 2018. A design pattern approach for multi-game level generation. In *Artificial Intelligence and Interactive Digital Entertainment*, AAAI. 105

Bellemare, M. G., Naddaf, Y., Veness, J., and Bowling, M. 2013. The arcade learning environment: An evaluation platform for general agents. *Journal of Artificial Intelligence Research*, 47(1):253–279. DOI: 10.1613/jair.3912 4, 84, 85, 146, 160

Bontrager, P., Khalifa, A., Mendes, A., and Togelius, J. 2016. Matching games and algorithms for general video game playing. In *12th Artificial Intelligence and Interactive Digital Entertainment Conference*. 66, 114

Braylan, A. and Miikkulainen, R. 2016. Object-model transfer in the general video game domain. In *12th Artificial Intelligence and Interactive Digital Entertainment Conference*. 92

Brockman, G., Cheung, V., Pettersson, L., Schneider, J., Schulman, J., Tang, J., and Zaremba, W. 2016. Openai gym. *ArXiv Preprint ArXiv:1606.01540*. 84, 146, 160

Browne, C. and Maire, F. 2010. Evolutionary game design. *IEEE Transactions on Computational Intelligence and AI in Games*, 2(1):1–16. DOI: 10.1109/tciaig.2010.2041928 4

Browne, C., Powley, E., Whitehouse, D., Lucas, S., Cowling, P., Rohlfshagen, P., Tavener, S., Pérez Liébana, D., Samothrakis, S., and Colton, S. 2012. A survey of Monte Carlo tree search methods. *IEEE Transactions on Computational Intelligence and AI in Games*, 4:1:1–43. 32, 33, 35

Buitinck, L., Louppe, G., Blondel, M., Pedregosa, F., Mueller, A., Grisel, O., Niculae, V., Prettenhofer, P., Gramfort, A., Grobler, J., Layton, R., VanderPlas, J., Joly, A., Holt, B., and

Varoquaux, G. 2013. API design for machine learning software: Experiences from the scikit-learn project. In *ECML PKDD Workshop: Languages for Data Mining and Machine Learning*, 108–122. 72

Chaslot, G. M. J.-B., Bakkes, S., Szita, I., and Spronck, P. 2006. Monte-Carlo tree search: A new framework for game AI. In *Proc. of the Artificial Intelligence for Interactive Digital Entertainment Conference*, 216–217. 32

Chu, C. Y., Harada, T., and Thawonmas, R. 2015. Biasing Monte-Carlo rollouts with potential field in general video game playing, 1–6. 65

Cook, M. and Colton, S. 2014. A rogue dream: Automatically generating meaningful content for games. In *10th Artificial Intelligence and Interactive Digital Entertainment Conference*. 115

Cook, M., Colton, S., Raad, A., and Gow, J. 2013. Mechanic miner: Reflection-driven game mechanic discovery and level design. In *European Conference on the Applications of Evolutionary Computation*, 284–293, Springer. DOI: 10.1007/978-3-642-37192-9_29 115

Coulom, R. 2006. Efficient selectivity and backup operators in Monte-Carlo tree search. In *Proc. of the 5th International Conference on Computer Games*, 72–83, Springer-Verlag. DOI: 10.1007/978-3-540-75538-8_7 32

Dahlskog, S. and Togelius, J. 2012. Patterns and procedural content generation: Revisiting Mario in world 1 level 1. In *Proc. of the 1st Workshop on Design Patterns in Games*, 1, ACM. DOI: 10.1145/2427116.2427117 106

Deepmind, G. 2019. AlphaStar. `https://deepmind.com/blog/alphastar-mastering-real-time-strategy-game-starcraft-ii/` 3

Dockhorn, A. and Apeldoorn, D. 2018. Forward model approximation for general video game learning. In *Computational Intelligence and Games (CIG), IEEE Conference on*. DOI: 10.1109/cig.2018.8490411 93

Ebner, M., Levine, J., Lucas, S. M., Schaul, T., Thompson, T., and Togelius, J. 2013. Towards a video game description language. *Dagstuhl Follow-Ups*, 6. 6

Finnsson, H. 2012. Generalized Monte-Carlo tree search extensions for general game playing. In *AAAI*. 32

Font, J. M., Mahlmann, T., Manrique, D., and Togelius, J. 2013. A card game description language. In *Applications of Evolutionary Computing, EvoApplications*, vol. 7835 of *LNCS*, 254–263, Vienna, Springer-Verlag. DOI: 10.1007/978-3-642-37192-9_26 4

Gaina, R. D., Couëtoux, A., Soemers, D. J., Winands, M. H., Vodopivec, T., Kirchgessner, F., Liu, J., Lucas, S. M., and Pérez Liébana, D. 2017a. The 2016 two-player GVGAI competition. *IEEE Transactions on Computational Intelligence and AI in Games*. DOI: 10.1109/tci-aig.2017.2771241 9, 20, 62

Gaina, R. D., Liu, J., Lucas, S. M., and Pérez Liébana, D. 2017b. Analysis of vanilla rolling horizon evolution parameters in general video game playing. In *European Conference on the Applications of Evolutionary Computation*, 418–434, Springer. DOI: 10.1007/978-3-319-55849-3_28 51, 67

Gaina, R. D., Lucas, S. M., and Pérez Liébana, D. 2017a. Population seeding techniques for rolling horizon evolution in general video game playing. In *Conference on Evolutionary Computation*, IEEE. DOI: 10.1109/cec.2017.7969540 52, 54, 67

Gaina, R. D., Lucas, S. M., and Pérez Liébana, D. 2017b. Rolling horizon evolution enhancements in general video game playing. In *IEEE Conference on Computational Intelligence and Games (CIG)*, IEEE. DOI: 10.1109/cig.2017.8080420 54, 55, 56, 67

Gaina, R. D., Lucas, S. M., and Pérez Liébana, D. 2018. General win prediction from agent experience. In *Computational Intelligence and Games (CIG), IEEE Conference on*. DOI: 10.1109/cig.2018.8490439 61

Gaina, R. D., Lucas, S. M., and Pérez Liébana, D. 2019. Tackling sparse rewards in real-time games with statistical forward planning methods. In *AAAI Conference on Artificial Intelligence (AAAI-19)*. 51, 55

Gaina, R. D., Pérez Liébana, D., and Lucas, S. M. 2016. General video game for 2 players: Framework and competition. In *IEEE Computer Science and Electronic Engineering Conference*. 9, 20

Gaina, R. D. 2016. The 2 player general video game playing competition. Master's thesis, University of Essex. 9

Gelly, S. and Silver, D. 2011. Monte-Carlo tree search and rapid action value estimation in computer Go. *Artificial Intelligence*, 175(11):1856–1875. DOI: 10.1016/j.artint.2011.03.007 32

Gelly, S., Wang, Y., Munos, R., and Teytaud, O. 2006. Modification of UCT with patterns in Monte-Carlo Go. *Technical Report*, Inst. Nat. Rech. Inform. Auto. (INRIA), Paris. 33

Genesereth, M., Love, N., and Pell, B. 2005. General game playing: Overview of the AAAI competition. *AI Magazine*, 26(2):62. 4

Glickman, M. E. 2012. Example of the Glicko-2 system. Boston University. 28

Gonzalez-Castro, J. M. and Pérez Liébana, D. 2017. Opponent models comparison for 2 players in GVGAI competitions. In *Computer Science and Electronic Engineering Conference, 9th*, IEEE. DOI: 10.1109/ceec.2017.8101616 159

Guckelsberger, C., Salge, C., Gow, J., and Cairns, P. 2017. Predicting player experience without the player: An exploratory study. In *Proc. of the Annual Symposium on Computer-Human Interaction in Play, CHI PLAY '17*, 305–315, New York, ACM. DOI: 10.1145/3116595.3116631 77

Guerrero-Romero, C., Lucas, S. M., and Pérez Liébana, D. 2018. Using a team of general AI algorithms to assist game design and testing. In *Conference on Computational Intelligence and Games (CIG)*. DOI: 10.1109/cig.2018.8490417 10

Hingston, P. 2010. A new design for a turing test for bots. In *Proc. of the IEEE Conference on Computational Intelligence in Games*, 345–350, IEEE. DOI: 10.1109/itw.2010.5593336 3

Hingston, P. 2012. *Believable Bots: Can Computers Play Like People?*, Springer. DOI: 10.1007/978-3-642-32323-2 11

Horn, H., Volz, V., Pérez Liébana, D., and Preuss, M. 2016. MCTS/EA hybrid GVGAI players and game difficulty estimation. In *Conference on Computational Intelligence in Games (CIG)*, 1–8, IEEE. DOI: 10.1109/cig.2016.7860384 56

Isaksen, A., Gopstein, D., Togelius, J., and Nealen, A. 2015. Discovering unique game variants. In *Computational Creativity and Games Workshop at the International Conference on Computational Creativity*. 118

Isaksen, A., Gopstein, D., and Nealen, A. 2015. Exploring game space using survival analysis. In *FDG*. 118

Jemerov, D. and Isakova, S. 2017. *Kotlin in Action*. Manning Publications. 145

Johnson, M., Hofmann, K., Hutton, T., and Bignell, D. 2016. The Malmo platform for artificial intelligence experimentation. In *IJCAI*, 4246–4247. 84, 160

Johnson, L., Yannakakis, G. N., and Togelius, J. 2010. Cellular automata for real-time generation of infinite cave levels. In *Proc. of the Workshop on Procedural Content Generation in Games*, 10, ACM. DOI: 10.1145/1814256.1814266 105

Joppen, T., Moneke, M. U., Schröder, N., Wirth, C., and Fürnkranz, J. 2018. Informed hybrid game tree search for general video game playing. *IEEE Transactions on Games*, 10(1):78–90. DOI: 10.1109/tciaig.2017.2722235 64, 66

Justesen, N., Bontrager, P., Togelius, J., and Risi, S. 2017. Deep learning for video game playing. *ArXiv Preprint ArXiv:1708.07902*. DOI: 10.1109/tg.2019.2896986 95

Justesen, N., Torrado, R. R., Bontrager, P., Khalifa, A., Togelius, J., and Risi, S. 2018. Illuminating generalization in deep reinforcement learning through procedural level generation. *ArXiv:1806.10729*. 13, 93

Justesen, N., Mahlmann, T., and Togelius, J. 2016. Online evolution for multi-action adversarial games. In *European Conference on the Applications of Evolutionary Computation*, 590–603, Springer. DOI: 10.1007/978-3-319-31204-0_38 49

Kazimipour, B., Li, X., and Qin, A. K. 2014. A review of population initialization techniques for evolutionary algorithms. In *Evolutionary Computation (CEC), IEEE Congress on*, 2585–2592, IEEE. DOI: 10.1109/cec.2014.6900618 52

Kempka, M., Wydmuch, M., Runc, G., Toczek, J., and Jaśkowski, W. 2016. Vizdoom: A doombased AI research platform for visual reinforcement learning. In *Conference on Computational Intelligence and Games*, 1–8, IEEE. DOI: 10.1109/cig.2016.7860433 3

Khalifa, A., Pérez Liébana, D., Lucas, S. M., and Togelius, J. 2016. General video game level generation. In *Proc. of the Genetic and Evolutionary Computation Conference*, 253–259, ACM. DOI: 10.1145/2908812.2908920 97, 98, 103

Khalifa, A., Green, M. C., Pérez Liébana, D., and Togelius, J. 2017. General video game rule generation. In *IEEE Conference on Computational Intelligence and Games (CIG)*. DOI: 10.1109/cig.2017.8080431 97, 109

Kim, K.-J., Choi, H., and Cho, S.-B. 2007. Hybrid of evolution and reinforcement learning for Othello players. In *Computational Intelligence and Games, CIG. IEEE Symposium on*, 203–209, IEEE. DOI: 10.1109/cig.2007.368099 52

Kimbrough, S. O., Koehler, G. J., Lu, M., and Wood, D. H. 2008. On a feasible—infeasible two-population (fi-2pop) genetic algorithm for constrained optimization: Distance tracing and no free lunch. *European Journal of Operational Research*, 190(2):310–327. DOI: 10.1016/j.ejor.2007.06.028 101, 113

Kocsis, L. and Szepesvári, C. 2006. Bandit based Monte-Carlo planning. *Machine Learning: ECML*, 4212:282–293. DOI: 10.1007/11871842_29 32, 33

Koster, R. 2013. *Theory of Fun for Game Design*. O'Reilly Media, Inc. 115

Kunanusont, K., Gaina, R. D., Liu, J., Pérez Liébana, D., and Lucas, S. M. 2017. The N-tuple bandit evolutionary algorithm for automatic game improvement. In *IEEE Proc. of the Congress on Evolutionary Computation (CEC)*, 2201–2208. DOI: 10.1109/cec.2017.7969571 118

Kunanusont, K., Lucas, S. M., and Pérez Liébana, D. 2017. General video game AI: Learning from screen capture. In *IEEE Conference on Evolutionary Computation (CEC)*, IEEE. DOI: 10.1109/cec.2017.7969556 92

Kunanusont, K., Lucas, S. M., and Pérez Liébana, D. 2018. Modelling player experience with the N-tuple bandit evolutionary algorithm. In *Artificial Intelligence and Interactive Digital Entertainment (AIIDE)*. 144

Kunanusont, K. 2016. General video game artificial intelligence: Learning from screen capture. Master's thesis, University of Essex. 9, 92

Lee, C.-S., Wang, M.-H., Chaslot, G. M. J.-B., Hoock, J.-B., Rimmel, A., Teytaud, O., Tsai, S.-R., Hsu, S.-C., and Hong, T.-P. 2009. The computational intelligence of MoGo revealed in Taiwan's computer Go tournaments. *IEEE Transactions on Computational Intelligence and AI in Games*, 1(1):73–89. DOI: 10.1109/tciaig.2009.2018703 32

Lehman, J. and Miikkulainen, R. 2015. General video game playing as a benchmark for human-competitive AI. In *AAAI-15 Workshop on Beyond the Turing Test*. 161

Levine, J., Congdon, C. B., Ebner, M., Kendall, G., Lucas, S. M., Miikkulainen, R., Schaul, T., and Thompson, T. 2013. General video game playing. *Dagstuhl Follow-Ups*, 6. 6, 9

Liu, J., Togelius, J., Pérez Liébana, D., and Lucas, S. M. 2017. Evolving game skill-depth using general video game AI agents. In *IEEE Proc. of the Congress on Evolutionary Computation (CEC)*. DOI: 10.1109/cec.2017.7969583 118

Liu, J., Pérez Liébana, D., and Lucas, S. M. 2016. Rolling horizon coevolutionary planning for two-player video games. In *Computer Science and Electronic Engineering (CEEC), 8th*, 174–179, IEEE. DOI: 10.1109/ceec.2016.7835909 49

Liu, J., Pérez Liébana, D., and Lucas, S. M. 2017a. Bandit-based random mutation hill-climbing. In *Evolutionary Computation (CEC), IEEE Congress on*, 2145–2151, IEEE. DOI: 10.1109/cec.2017.7969564 54

Liu, J., Pérez Liébana, D., and Lucas, S. M. 2017b. The single-player GVGAI learning framework—technical manual. 81

Liu, J. 2017. GVGAI single-player learning competition. *IEEE CIG17*. 88

Loiacono, D., Lanzi, P. L., Togelius, J., Onieva, E., Pelta, D. A., Butz, M. V., Lonneker, T. D., Cardamone, L., Pérez Liébana, D., Sáez, Y., et al. 2010. The 2009 simulated car racing championship. *IEEE Transactions on Computational Intelligence and AI in Games*, 2(2):131–147. DOI: 10.1109/tciaig.2010.2050590 3

Love, N., Hinrichs, T., Haley, D., Schkufza, E., and Genesereth, M. 2008. General game playing: Game description language specification. 4

Lucas, S. M., Liu, J., and Pérez Liébana, D. 2018a. The N-tuple bandit evolutionary algorithm for game agent optimisation. *ArXiv Preprint ArXiv:1802.05991*. DOI: 10.1109/cec.2018.8477869 155

Lucas, S. M., Liu, J., and Pérez Liébana, D. 2018b. The N-tuple bandit evolutionary algorithm for game agent optimisation. *ArXiv Preprint ArXiv:1802.05991*. DOI: 10.1109/cec.2018.8477869 49, 120, 121

Lucas, S. M., Liu, J., and Pérez Liébana, D. 2018c. The N-tuple bandit evolutionary algorithm for game agent optimisation. *ArXiv Preprint ArXiv:1802.05991*. DOI: 10.1109/cec.2018.8477869 118

Lucas, S. M., Samothrakis, S., and Pérez Liébana, D. 2014. Fast evolutionary adaptation for Monte Carlo tree search. In *Proc. of EvoGames*, to appear. DOI: 10.1007/978-3-662-45523-4_29 36, 39

Lucas, S. M. 2018. Game AI research with fast planet wars variants. In *IEEE Conference on Computational Intelligence and Games*. DOI: 10.1109/cig.2018.8490377 148, 152

Miettinen, K. 1999. *Non-Linear Multiobjective Optimization*. Kluwer, International Series in Operations Research and Management Science. DOI: 10.1007/978-1-4615-5563-6 59

Mnih, V., Kavukcuoglu, K., Silver, D., Rusu, A. A., Veness, J., Bellemare, M. G., Graves, A., Riedmiller, M., Fidjeland, A. K., Ostrovski, G., et al. 2015. Human-level control through deep reinforcement learning. *Nature*, 518(7540):529. DOI: 10.1038/nature14236 4

Myerson, R. B. 1991. *Game Theory: Analysis of Conflict*. Cambridge (MA), London, Harvard University Press, 1997. DOI: 10.2307/j.ctvjsf522 45

Nelson, M. J., Togelius, J., Browne, C., and Cook, M. 2014. *Rules and Mechanics*. In Shaker, N., Togelius, J., and Nelson, M. J., Eds., *Procedural Content Generation in Games: A Textbook and an Overview of Current Research*, 97–117, Springer. 4

Nelson, M. J. 2016. Investigating vanilla MCTS scaling on the GVG-AI game corpus. In *IEEE Conference on Computational Intelligence and Games (CIG)*, 402–408, IEEE. DOI: 10.1109/cig.2016.7860443 66

Neufeld, X., Mostaghim, S., and Pérez Liébana, D. 2015. Procedural level generation with answer set programming for general video game playing. In *7th Computer Science and Electronic Engineering Conference (CEEC)*, 207–212, IEEE. DOI: 10.1109/ceec.2015.7332726 9, 107

Neufeld, X. 2016. Procedural level generation with answer set programming for general video game playing. Master's thesis, University of Magdeburg. DOI: 10.1109/ceec.2015.7332726 9

Newborn, M. 2003. *Computer Chess*. John Wiley & Sons Ltd. 3

Nichols, J. 2016. The use of genetic algorithms in automatic level generation. Master's thesis, University of Essex. 9, 97, 106

Nielsen, T. S., Barros, G. A., Togelius, J., and Nelson, M. J. 2015. General video game evaluation using relative algorithm performance profiles. In *European Conference on the Applications of Evolutionary Computation*, 369–380, Springer. DOI: 10.1007/978-3-319-16549-3_30 102, 113, 115

Ontanon, S., Synnaeve, G., Uriarte, A., Richoux, F., Churchill, D., and Preuss, M. 2013. A survey of real-time strategy game AI research and competition in StarCraft. *IEEE Transactions on Computational Intelligence and AI in Games*, 5(4):293–311. DOI: 10.1109/tci-aig.2013.2286295 3

Pérez Liébana, D., Samothrakis, S., Lucas, S., and Rohlfshagen, P. 2013. Rolling horizon evolution versus tree search for navigation in single-player real-time games. In *Proc. of the 15th Annual Conference on Genetic and Evolutionary Computation*, 351–358, ACM. DOI: 10.1145/2463372.2463413 49

Pérez Liébana, D., Samothrakis, S., Togelius, J., Schaul, T., and Lucas, S. 2014. The general video game AI competition. www.gvgai.net 42

Pércz Liébana, D., Samothrakis, S., Togelius, J., Schaul, T., Lucas, S., Couëtoux, A., Lee, J., Lim, C.-U., and Thompson, T. 2015. The 2014 general video game playing competition. *IEEE Transactions on Computational Intelligence and AI in Games*, 8:229–243. DOI: 10.1109/tciaig.2015.2402393 63

Pérez Liébana, D., Dieskau, J., Hunermund, M., Mostaghim, S., and Lucas, S. 2015a. Open loop search for general video game playing. In *Proc. of the Annual Conference on Genetic and Evolutionary Computation*, 337–344, ACM. DOI: 10.1145/2739480.2754811 55, 56

Pérez Liébana, D., Dieskau, J., Hunermund, M., Mostaghim, S., and Lucas, S. M. 2015b. Open loop search for general video game playing. In *Proc. of the on Genetic and Evolutionary Computation Conference, (GECCO)*, 337–344, Association for Computing Machinery (ACM). DOI: 10.1145/2739480.2754811 43

Pérez Liébana, D., Samothrakis, S., Togelius, J., Schaul, T., Lucas, S. M., Couëtoux, A., Lee, J., Lim, C.-U., and Thompson, T. 2016. The 2014 general video game playing competition. *IEEE Transactions on Computational Intelligence and AI in Games*, 8(3):229–243. DOI: 10.1109/tciaig.2015.2402393 102

Pérez Liébana, D., Stephenson, M., Gaina, R. D., Renz, J., and Lucas, S. M. 2017. Introducing real world physics and macro-actions to general video game AI. In *Conference on Computational Intelligence and Games (CIG)*, IEEE. DOI: 10.1109/cig.2017.8080443 19, 66

Pérez Liébana, D., Hofmann, K., Mohanty, S. P., Kuno, N., Kramer, A., Devlin, S., Gaina, R. D., and Ionita, D. 2018a. The multi-agent reinforcement learning in MalmÖ (MARLÖ) competition. In *Challenges in Machine Learning (CiML, NIPS Workshop)*, 1–4. 5

Pérez Liébana, D., Liu, J., Khalifa, A., Gaina, R. D., Togelius, J., and Lucas, S. M. 2018b. General video game AI: A multi-track framework for evaluating agents, games and content generation algorithms. *ArXiv Preprint ArXiv:1802.10363*. DOI: 10.1109/tg.2019.2901021 65, 159

Pérez Liébana, D., Mostaghim, S., and Lucas, S. M. 2016. Multi-objective tree search approaches for general video game playing. In *Congress on Evolutionary Computation (CEC)*, 624–631, IEEE. DOI: 10.1109/cec.2016.7743851 44, 45, 48

Pérez Liébana, D., Rohlfshagen, P., and Lucas, S. M. 2012. The physical travelling salesman problem: WCCI 2012 competition. In *Proc. of the IEEE Congress on Evolutionary Computation*, 1–8, IEEE. DOI: 10.1109/cec.2012.6256440 3

Pérez Liébana, D., Samothrakis, S., and Lucas, S. 2014. Knowledge-based fast evolutionary MCTS for general video game playing. In *Conference on Computational Intelligence and Games*, 1–8, IEEE. DOI: 10.1109/cig.2014.6932868 66

Philipp Rohlfshagen, Jialin Liu, D. P.-L., and Lucas, S. M. 2017. Pac-Man conquers academia: Two decades of research using a classic arcade game. *IEEE Transactions on Computational Intelligence and AI in Games*. DOI: 10.1109/tg.2017.2737145 160

Prada, R., Melo, F., and Quiterio, J. 2014. Geometry friends competition. 3

Rohlfshagen, P. and Lucas, S. M. 2011. Ms Pac-Man vs. ghost team CEC 2011 competition. In *Proc. of the IEEE Congress on Evolutionary Computation*, 70–77, IEEE. DOI: 10.1109/cec.2011.5949599 3

Ross, B. 2014. General video game playing with goal orientation. Master's thesis, University of Strathclyde. 9

Russell, S. J. and Norvig, P. 2016. *Artificial Intelligence: A Modern Approach*. Malaysia, Pearson Education Limited. 91, 92

Samothrakis, S., Pérez Liébana, D., Lucas, S. M., and Fasli, M. 2015. Neuroevolution for general video game playing. In *Conference on Computational Intelligence and Games (CIG)*, 200–207, IEEE. DOI: 10.1109/cig.2015.7317943 92

Santos, B. S., Bernardino, H. S., and Hauck, E. An improved rolling horizon evolution algorithm with shift buffer for general game playing. In *17th Brazilian Symposium on Computer Games and Digital Entertainment (SBGames)*, 382–388, IEEE, 2018. 49, 56

Schaul, T., Togelius, J., and Schmidhuber, J. 2011. Measuring intelligence through games. *CoRR* abs/1109.1314:1–19. 4

Schaul, T.   2013a.   A video game description language for model-based or interactive learning.   In *IEEE Conference on Computational Intelligence in Games (CIG)*, 1–8. DOI: 10.1109/cig.2013.6633610 6, 13

Schaul, T. 2013b. A video game description language for model-based or interactive learning. In *Proc. of the IEEE Conference on Computational Intelligence in Games*, 193–200, Niagara Falls, IEEE Press. DOI: 10.1109/cig.2013.6633610 22

Schaul, T.   2014.   An extensible description language for video games.   *IEEE Transactions on Computational Intelligence and AI in Games*, 6(4):325–331. DOI: 10.1109/tci-aig.2014.2352795 6

Schuster, T. 2015. MCTS based agent for general video games. Master's thesis, Maastricht University. 9

Shaker, N., Liapis, A., Togelius, J., Lopes, R., and Bidarra, R. 2016. Constructive generation methods for dungeons and levels. In *Procedural Content Generation in Games*, 31–55, Springer. DOI: 10.1007/978-3-319-42716-4_3 97

Shaker, N., Togelius, J., and Nelson, M. J.   2016.   *Procedural Content Generation in Games*, Springer. DOI: 10.1007/978-3-319-42716-4 11, 97, 116

Sharif, M., Zafar, A., and Muhammad, U. 2017. Design patterns and general video game level generation. *International Journal of Advanced Computer Science and Applications*, 8(9):393–398. DOI: 10.14569/ijacsa.2017.080952 106

Silver, D., Huang, A., Maddison, C. J., Guez, A., Sifre, L., Van Den Driessche, G., Schrittwieser, J., Antonoglou, I., Panneershelvam, V., Lanctot, M., et al.   2016.   Mastering the game of Go with deep neural networks and tree search. *Nature*, 529(7587):484–489. DOI: 10.1038/nature16961 3

Soemers, D. 2016. Enhancements for real-time Monte-Carlo tree search in general video game playing. Master's thesis, Maastricht University. DOI: 10.1109/cig.2016.7860448 9

Stephenson, M., Anderson, D., Khalifa, A., Levine, J., Renz, J., Togelius, J., and Salge, C. 2018. A continuous information gain measure to find the most discriminatory problems for AI benchmarking. *ArXiv Preprint ArXiv:1809.02904*. 108

Sutton, R. S. and Barto, A. G. 1998. *Reinforcement Learning: An Introduction*, MIT Press. DOI: 10.1109/tnn.1998.712192 62

Togelius, J. and Schmidhuber, J.   2008.   An experiment in automatic game design.   In *Proc. of the IEEE Symposium on Computational Intelligence and Games (CIG)*, 111–118. DOI: 10.1109/cig.2008.5035629 115

Togelius, J., Shaker, N., Karakovskiy, S., and Yannakakis, G. N. 2013. The Mario AI championship 2009–2012. *AI Magazine*, 34(3):89–92. DOI: 10.1609/aimag.v34i3.2492 3

Torrado, R. R., Bontrager, P., Togelius, J., Liu, J., and Pérez Liébana, D. 2018. Deep reinforcement learning in the general video game AI framework. In *IEEE Conference on Computational Intelligence and Games (CIG)*. DOI: 10.1109/cig.2018.8490422 25, 83, 90, 91, 93, 146

Turing, A. M. 1953. *Chess*. In Bowden, B. V., Ed., *Faster than Thought*, 286–295, Pitman. 3

Turk, G. 1991. Generating textures on arbitrary surfaces using reaction-diffusion. In *ACM SIGGRAPH Computer Graphics*, vol. 25, 289–298, ACM. DOI: 10.1145/127719.122749 97

van Eeden, J. 2015. Analysing and improving the knowledge-based fast evolutionary MCTS algorithm. Master's thesis. 9

Vinyals, O., Ewalds, T., Bartunov, S., Georgiev, P., Vezhnevets, A. S., Yeo, M., Makhzani, A., Küttler, H., Agapiou, J., Schrittwieser, J., et al. 2017. Starcraft II: A new challenge for reinforcement learning. *ArXiv Preprint ArXiv:1708.04782*. 3

Vodopivec, T., Samothrakis, S., and Ster, B. 2017. On Monte Carlo tree search and reinforcement learning. *Journal of Artificial Intelligence Research*, 60:881–936. DOI: 10.1613/jair.5507 64

Volz, V., Ashlock, D., Colton, S., Dahlskog, S., Liu, J., Lucas, S., Pérez Liébana, D., and Thompson, T. 2017. Gameplay evaluation measures. *Dagstuhl Follow-Ups*. 160

Volz, V., Ashlock, D., Colton, S., Dahlskog, S., Liu, J., Lucas, S. M., Pérez Liébana, D., and Thompson, T. 2018. *Gameplay Evaluation Measures*. In André, E., Cook, M., Preuß, M., and Spronck, P., Eds., *Artificial and Computational Intelligence in Games: AI-Driven Game Design (Dagstuhl Seminar 17471)*, 36–39, Dagstuhl, Germany, Schloss Dagstuhl–Leibniz-Zentrum fuer Informatik. 77

Waard, M. d., Roijers, D. M., and Bakkes, S. C. 2016. Monte Carlo tree search with options for general video game playing. In *IEEE Conference on Computational Intelligence and Games (CIG)*, 47–54, IEEE. DOI: 10.1109/cig.2016.7860383 9, 66

Weinstein, A. and Littman, M. L. 2012. Bandit-based planning and learning in continuous-action Markov decision processes. In *Proc. of the 22nd International Conference on Automated Planning and Scheduling, ICAPS*, Brazil. 62

Weiss, K., Khoshgoftaar, T. M., and Wang, D. 2016. A survey of transfer learning. *Journal of Big Data*, 3(1):9. DOI: 10.1186/s40537-016-0043-6 5

Williams, P. R., Pérez Liébana, D., and Lucas, S. M. 2016. Ms. Pac-Man vs. ghost team CIG 2016 competition. In *Computational Intelligence and Games (CIG), 2016 IEEE Conference on*. DOI: 10.1109/cig.2016.7860446 160

Zhu, J., Rosset, S., Zou, H., and Hastie, T. 2006. Multi-class AdaBoost. *Ann Arbor*, 1001(48109):1612. DOI: 10.4310/sii.2009.v2.n3.a8 72

Zitzler, E. 1999. *Evolutionary Algorithms for Multiobjective Optimization: Methods and Applications*. TIK-Schriftenreihe Nr. 30, Diss ETH No. 13398, Swiss Federal Institute of Technology (ETH) Zurich: Shaker Verlag, Germany. 42

Authors' Biographies

DIEGO PÉREZ LIÉBANA

Diego Pérez Liébana is a Lecturer in Computer Games and AI at Queen Mary University of London (QMUL) and holds a Ph.D. in Computer Science from the University of Essex (2015). His research interests are search algorithms, evolutionary computation, and reinforcement learning applied to real-time games and general video game playing. He's published more than 60 papers in leading conferences and journals in the area, including best paper awards (CIG, EvoStar). He's the main organizer behind popular AI game-based competitions in the field, serves as a reviewer in top conferences and journals, and he was general chair of the 2019 IEEE Conference on Games (QMUL). He has experience in the video games industry with titles published for both PC and consoles, and also in developing AI tools for games.

SIMON M. LUCAS

Simon M. Lucas is a professor of Computer Science at Queen Mary University of London (UK) where he is the Head of School and leads the Game Artificial Intelligence Group. He holds a Ph.D. (1991) in Electronics and Computer Science from the University of Southampton. His main research interests are games, evolutionary computation, and machine learning, and he has published widely in these fields with over 180 peer-reviewed papers. He is the inventor of the scanning n-tuple classifier and is the founding Editor-in-Chief of the *IEEE Transactions on Computational Intelligence* and *AI in Games*.

RALUCA D. GAINA

Raluca D. Gaina is currently studying for her Ph.D. in Intelligent Games and Games Intelligence at Queen Mary University of London, in the area of rolling horizon evolution in general video game playing, after completing a B.Sc. and M.Sc. in Computer Games at the University of Essex. In 2018, she did a three-month internship at Microsoft Research Cambridge, working on the Multi-Agent Reinforcement Learning in Malmo Competition (MARLO, aka.ms/marlo). She is the track organizer of the Two-Player General Video Game AI Competition (gvgai.net). Her research interests include general video game playing AI, reinforcement learning, and evolutionary computation algorithms.

JULIAN TOGELIUS

Julian Togelius is an Associate Professor in the Department of Computer Science and Engineering, New York University, USA. He works on artificial intelligence for games and games for artificial intelligence. His current main research directions involve search-based procedural content generation in games, general video game playing, player modeling, generating games based on open data, and fair and relevant benchmarking of AI through game-based competitions. He is the Editor-in-Chief of *IEEE Transactions on Games* and has been chair or program chair of several of the main conferences on AI and games. Togelius holds a B.A. from Lund University, an M.Sc. from the University of Sussex, and a Ph.D. from the University of Essex. He has previously worked at IDSIA in Lugano and at the IT University of Copenhagen.

AHMED KHALIFA

Ahmed Khalifa is a Ph.D. student at New York University working on procedural content generation and automated game playing. He also works as a game developer and designer in his free time and has released more than 30 games algorithms.

JIALIN LIU

Jialin Liu is currently a Research Assistant Professor at the Department of Computer Science and Engineering of Southern University of Science and Technology (SUSTech, China). Before joining SUSTech, she was a Postdoctoral Research Associate at Queen Mary University of London (QMUL, UK) and one of the founding members of the Game AI research group of QMUL. Her research interests include AI and games, noisy optimization, portfolio of algorithms, and meta-heuristics. Jialin serves as Program Co-Chair of 2018 IEEE Computational Intelligence and Games (CIG2018), and Competition Chair of FDG2018, FDG2019, and CEC2019. Jialin received her Ph.D. in Computer Science from the Inria Saclay and the Université Paris-Saclay (France) in December 2015 and a Master's in Bioinformatics and Biostatistics from the École Polytechnique and the Université Paris-Sud (France) in 2013.

Printed in the United States
by Baker & Taylor Publisher Services